Skills Worksheet

Concept Review

Section: States and State Changes

Complete each statement below by choosing a term from the following list. Use each term only once.

solid	cohesion	melting	surface tension
liquid	adhesion	evaporation	boiling point
gas	deposition	condensation	melting point
viscous	freezing	sublimation	freezing point

1. The particles in a _____ are very close together in an orderly, fixed, and usually crystalline arrangement. _____ is an endo-thermic change of state in which a solid becomes a liquid. The temperature and pressure at which a solid becomes a liquid is its _____.

2. Because particles in a _____ have enough kinetic energy to be able to move past each other easily, they take the shape of their container. While many liquids flow readily, many are resistant to flowing, or are

   _____.

3. Because they are held close together, liquid particles are more affected by forces between particles. They have attraction for each other, or

   _____, as well as attraction for particles of solid surfaces, called

   _____. Liquids tend to form spherical drops because of

   _____, or the tendency to decrease their surface area to the smallest size possible, thereby decreasing their energy. Particles in a liquid can gain enough kinetic energy to leave the surface and become a gas in a

   process called _____.

4. Attractive forces between _____ particles do not have a great effect, which makes the particles essentially independent of each other. The temperature and pressure at which the number of liquid particles becoming gas particles is the same as the number of gas particles returning to the liquid

   phase is called a substance's _____. Gas particles lose energy and

   become liquid during _____.

5. The process during which a liquid substance loses energy and becomes a

solid is called _____. The temperature at which this change

occurs is the _____ for a substance.

6. The particles of solids may become gas particles without first melting in a

process called _____. The reverse of this process, in which a gas

becomes a solid without first becoming liquid, is called _____.

Skills Worksheet

Concept Review

Section: Intermolecular Forces

Write the answers to the following questions in the space provided.

1. Why do ionic compounds tend to have higher boiling and melting points than molecular compounds have?

   _____

   _____

   _____

2. Why do molecular substances with weak intermolecular forces have low melting points?

   _____

   _____

   _____

3. Why do molecular substances with strong intermolecular forces have high melting points?

   _____

   _____

   _____

4. How do dipole-dipole forces affect the melting and boiling points of substances?

   _____

   _____

   _____

   _____

Concept Review continued

5. What forces are involved in hydrogen bonding?

_____

_____

_____

_____

6. What effect does hydrogen bonding have on the physical properties of water?

_____

_____

_____

_____

_____

7. How can a molecule have a momentary dipole?

_____

_____

_____

8. Name the force of attraction between molecules with momentary dipoles.

_____

9. How do London forces and dipole-dipole forces between molecules differ from forces between ions in crystals?

_____

_____

_____

10. Explain the role of particle size and shape on the strength of attractive forces.

_____

_____

_____

_____

Skills Worksheet

Concept Review

Section: Energy of State Changes

Complete each statement below by writing the correct word or words in the space provided. Refer to Figure 1, below, to answer items 1–6.

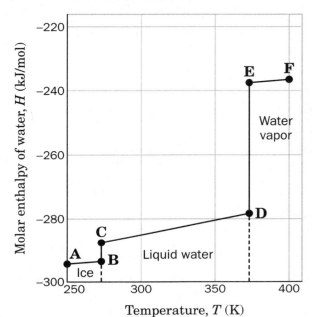

Figure 1

1. The molar enthalpy change at 273.15 K (B → C) is called the

_____

2. The molar enthalpy change at 373.15 K (D → E) is called the

_____

3. The slopes of the slowly rising lines in the graph, A → B, C → D, and E → F,

are the _____ of water at each state.

4. The large heat capacity of liquid water is attributed to _____

Solve the following problem and write your answer in the space provided.

5. The enthalpy of vaporization for nitric acid is 30.30 kJ/mol, and the molar entropy of vaporization is 84.4 J/mol·K. Calculate the boiling point of nitric acid.

▎Concept Review *continued*

Complete each statement below by circling the correct word or phrase in brackets.

6. It takes a lot [more, less] heat to vaporize the water than to melt ice.

7. Heat must be [absorbed, released] to raise the temperature of ice, water, or water vapor.

8. The enthalpy changes accompanying a change of state are much [smaller, greater] than those that accompany the heating of the substance at each state.

9. The tendency to [lower, higher] energy is seen in thermodynamics as positive ΔH values, and a(n) [increase, decrease] in disorder is seen as negative ΔS values.

10. During changes of state, changes in enthalpy and entropy [oppose, complement] each other. The relative values of ΔH and $T\Delta S$ determine which state is preferred.

11. During evaporation, a liquid becomes a gas at a temperature [at, well below] its boiling point.

12. Evaporation is a [endothermic, exothermic] process.

13. Gibbs energy relates entropy and enthalpy changes to the [spontaneity, rate] of a phase transition.

14. A process is spontaneous if ΔG is [positive, negative].

15. All spontaneous processes occur with a(n) [increase, decrease] in Gibbs energy.

16. When ΔG is exactly zero, the system is in a state of [flux, equilibrium].

17. The melting point of a solid equals the enthalpy of fusion [divided, multiplied] by the entropy of fusion.

18. Boiling occurs when the tendency toward [order, disorder] overcomes the tendency to lose energy.

19. [Condensation, evaporation] occurs when the tendency to lose energy overcomes the tendency to increase disorder.

20. Boiling points are pressure dependent because pressure has [no, a large] effect on the entropy of a gas.

Skills Worksheet

Concept Review

Section: Phase Equilibrium

Complete each statement below by choosing a term from the following list. Use each term only once.

phase	triple point	vapor pressure	normal boiling point
equilibrium	critical point	phase diagram	supercritical fluid

1. A _____ is a region that has the same composition and properties throughout. When particles are constantly moving between two or more phases, yet no net change in the amount of substance in either phase occurs,

   the system is said to be in _____. The pressure exerted by the molecules of a gas, or vapor, in equilibrium with a liquid is called the

   _____. When you increase a system's temperature to the point at which the vapor pressure of a substance is equal to standard atmospheric

   pressure, you have reached the substance's _____.

2. A graph of the relationship between the physical state of a substance and the

   temperature and pressure of the substance is called a _____. The temperature and pressure conditions at which the solid, liquid, and gaseous

   phases of a substance coexist at equilibrium is called the _____. The temperature and pressure at which the gas and liquid states of a

   substance become identical and form one phase is the _____.

   Above this temperature, the substance is referred to as a _____, and the liquid and vapor phases are indistinguishable.

Write the answer to the following questions in the space provided.

3. What physical factor does the average kinetic energy of molecules depend on?

   _____

   _____

4. Explain why the vapor pressure of molecules doubles or triples for every 10°C increase in temperature, while the kinetic energy increases only about 3%.

_____

_____

_____

5. Use the phase diagram for water in your text to complete the table below and items 6–8.

Description of point	Temperature	Pressure	Point
The temperature and pressure at which three phases of water exist in equilibrium			
The temperature at which water boils at 1.0 atm of pressure			
The temperature at which water freezes/melts at 1.0 atm of pressure			
The temperature and pressure at and above which the properties of water vapor cannot be distinguished from those of liquid water—water exists as a single phase			

Write the answer to the following questions in the space provided.

6. Name the phases that water will exhibit if the pressure is kept constant at 110 kPa and the temperature is gradually increased from −10°C to 110°C.

_____

7. Name the phases that water will exhibit if the pressure is kept constant at 0.31 kPa and the temperature is gradually increased from −10°C to 110°C. What term is given to the phase transformation of water that occurs under these conditions?

_____

8. Along which line segment do solids and liquids coexist? Describe the slope of this line for water. What will an increase in pressure do to water's melting point?

_____

_____

_____

Assessment

Quiz

Section: States and State Changes

In the space provided, write the letter of the term or phrase that best answers the question.

_____ **1.** A sample of matter whose particles are close together and cannot move past each other is
   a. a solid.
   b. a liquid.
   c. a gas.
   d. viscous.

_____ **2.** Fluids are materials that will flow from one place to another. Which of the following is a fluid?
   a. solid
   b. liquid
   c. gas
   d. Both (b) and (c)

_____ **3.** If particles have little attraction for each other and can freely move throughout the container, the particles are part of a
   a. solid.
   b. liquid.
   c. gas.
   d. None of the above

_____ **4.** The attraction of particles for each other within a liquid is
   a. adhesion.
   b. cohesion.
   c. surface tension.
   d. capillary action.

_____ **5.** Capillary action is the result of
   a. adhesion.
   b. cohesion.
   c. surface tension.
   d. Both (a) and (b)

_____ **6.** Water forms drops because of
   a. viscosity.
   b. capillary action.
   c. surface tension.
   d. fluidity.

Quiz *continued*

_____ **7.** A liquid becomes a gas during
    a. evaporation.
    b. condensation.
    c. sublimation.
    d. deposition.

_____ **8.** Some products advertise that they are "freeze-dried," which means
that water is removed from them while they are solids. What process
is used in producing these products?
    a. condensation
    b. evaporation
    c. sublimation
    d. deposition

_____ **9.** All changes of state are
    a. physical changes.
    b. chemical changes.
    c. evaporation.
    d. endothermic.

_____ **10.** Which of the following processes might occur when an object is
heated?
    a. condensation
    b. melting
    c. deposition
    d. freezing

Assessment

Quiz

Section: Intermolecular Forces

In the space provided, write the letter of the term or phrase that best answers the question.

_____ 1. Which of the following is most likely to have a high boiling point?
   a. NaCl
   b. CO_2
   c. CH_4
   d. H_2O

_____ 2. An ionic compound that contains a large ion with a 2+ charge will have ____ melting point compared to one containing a small ion with a 2+ charge, assuming the negative ions are the same.
   a. a higher
   b. a lower
   c. the same
   d. The two melting points cannot be compared.

_____ 3. Which of the following statements is true regarding intermolecular forces?
   a. They act only between oppositely charged ions.
   b. They are stronger than chemical bonds.
   c. They act over short distances.
   d. They increase as molecules get farther apart.

_____ 4. A dipole-dipole force is strongest when the molecules are
   a. far apart.
   b. nonpolar.
   c. strongly polar.
   d. large.

_____ 5. As dipole-dipole forces increase, melting points
   a. increase.
   b. decrease.
   c. remain the same.
   d. cannot be predicted.

_____ 6. Water's relatively high boiling point is the result of
   a. covalent bonding.
   b. hydrogen bonding.
   c. ionic bonding.
   d. London forces.

Quiz *continued*

_____ **7.** Hydrogen bonding is a special type of
   a. covalent bond.
   b. dipole-dipole force.
   c. ionic bond.
   d. London force.

_____ **8.** Which of the following will form the strongest hydrogen bonds?
   a. HF
   b. HCl
   c. H_2S
   d. HBr

_____ **9.** London dispersion forces exist as a result of
   a. attractions between ions.
   b. dipole-dipole attractions in polar molecules.
   c. the sharing of electrons.
   d. temporary dipoles in nonpolar molecules.

_____ **10.** The properties of a substance do *not* depend on the
   a. type of intermolecular force.
   b. size of its particles.
   c. shape of its particles.
   d. number of ions the particle contains.

Name _____ Class _____ Date _____

Quiz

Section: Energy of State Changes

In the space provided, write the letter of the term or phrase that best answers the question.

_____ **1.** The total energy content of a system is its
   a. entropy.
   b. enthalpy.
   c. temperature.
   d. free energy.

_____ **2.** The amount of energy needed to melt one mole of a substance is its molar
   a. enthalpy of fusion.
   b. enthalpy of vaporization.
   c. entropy of fusion.
   d. entropy of vaporization.

_____ **3.** A spontaneous change is favored by a change in entropy that is
   a. positive.
   b. negative.
   c. equal to its temperature change.
   d. equal to its enthalpy change.

_____ **4.** The symbol for a change in enthalpy is
   a. ΔG.
   b. ΔH.
   c. ΔS.
   d. ΔT.

_____ **5.** A system is at equilibrium, with water freezing at the same rate as it is melting. Which of the following is true?
   a. $\Delta H = \Delta S$
   b. $\Delta H > T\Delta S$
   c. $\Delta H = T\Delta S$
   d. $\Delta H < T\Delta S$

_____ **6.** If a change occurs spontaneously, which of the following is true?
   a. $\Delta H = \Delta S$
   b. $\Delta H > T\Delta S$
   c. $\Delta H = T\Delta S$
   d. $\Delta H < T\Delta S$

Quiz *continued*

_____ **7.** When $\Delta H_{vap} > T\Delta S_{vap}$, the state that is favored is
   a. solid.
   b. liquid.
   c. gaseous.
   d. equilibrium.

_____ **8.** If $\Delta S_{fus} = 20$ J/mol·K and $\Delta H_{fus} = 3000$ J/mol for a certain substance, its melting point is
   a. 0.007 K.
   b. 15 K.
   c. 150 K.
   d. 60 000 K.

_____ **9.** Pressure can affect the ____ of a liquid.
   a. boiling point
   b. freezing point
   c. entropy
   d. enthalpy

_____ **10.** Boiling points are affected by pressure changes because pressure changes
   a. affect the temperature of the gas.
   b. have a large effect on the entropy of a gas.
   c. cause a liquid to be compressed.
   d. change the melting point of the substance.

Assessment

Quiz

Section: Phase Equilibrium

In the space provided, write the letter of the term or phrase that best answers the question.

_____ **1.** Which of the following best describes a phase?
  a. It has the same composition throughout.
  b. It has the same properties throughout.
  c. It has the same properties and composition throughout.
  d. It has the same temperature throughout.

_____ **2.** Dry ice sublimes into carbon dioxide gas. If other conditions remain constant and nothing leaves the system, the dry ice and the carbon dioxide gas will
  a. become the same phase.
  b. reach equilibrium.
  c. stop changing to other phases.
  d. Both (a) and (b)

_____ **3.** Which of the following involves more than two phases?
  a. saltwater
  b. sweetened tea over ice
  c. oil and vinegar salad dressing
  d. a carbonated beverage over ice

_____ **4.** When temperature increases, vapor pressure
  a. decreases.
  b. increases.
  c. does not change.
  d. can either increase or decrease.

_____ **5.** The vapor pressure of water is 4.25 kPa at 30.0°C. The vapor pressure of water at 10.0°C might be
  a. −2.34 kPa.
  b. 1.23 kPa.
  c. 4.25 kPa.
  d. 12.34 kPa.

_____ **6.** The temperature at which the vapor pressure of a substance equals atmospheric pressure is its
  a. boiling point.
  b. critical point.
  c. melting point.
  d. triple point.

_____ **7.** The critical point of each substance is *not* a point at which
  a. pressure and temperature are greater than they are at the boiling point.
  b. all equilibrium lines meet on a phase diagram.
  c. the liquid and vapor phases are indistinguishable.
  d. any higher temperature results in a supercritical fluid.

_____ **8.** Phase diagrams can be used to identify substances because
  a. the triple point for every substance is unique.
  b. the basic structure of the diagram varies for each substance.
  c. supercritical fluids only appear for certain substances.
  d. liquids do not exist for all substances.

_____ **9.** When drawing a phase diagram, you need to know the relative densities of the solid and liquid and
  a. the vapor pressure of the solid or liquid at 1 atm pressure.
  b. the triple point.
  c. the critical point.
  d. All of the above

_____ **10.** On a phase diagram, assume you have point M on the liquid-vapor equilibrium line. If you raise the temperature slightly and keep the same pressure, the substance will now be
  a. a solid.
  b. a liquid.
  c. a gas.
  d. on a different equilibrium line.

Assessment

Chapter Test

States of Matter and Intermolecular Forces

In the space provided, write the letter of the term or phrase that best completes each statement or best answers each question.

_____ 1. If the particles in a sample of matter are attracted to each other but can move past each other, the matter is a
   a. solid.
   b. liquid.
   c. gas.
   d. plasma.

_____ 2. A gas does not have a definite volume because its particles are
   a. essentially independent.
   b. cohesive.
   c. adhesive.
   d. vibrating in place.

_____ 3. Which of the following is *not* an explanation of why water rises through the tissues in the stem of a plant?
   a. capillary action
   b. surface tension
   c. cohesion
   d. adhesion

_____ 4. Which of the following is true for the melting and freezing points of a pure substance?
   a. The melting point is higher than the freezing point.
   b. The melting point is lower than the freezing point.
   c. The melting point is the same as the freezing point.
   d. There is no relation between the freezing point and the melting point.

_____ 5. Condensation involves changing from a
   a. solid to a gas.
   b. liquid to a gas.
   c. gas to a liquid.
   d. gas to a solid.

_____ 6. In general, ionic compounds have
   a. high boiling points and low melting points.
   b. low boiling points and high melting points.
   c. high boiling points and high melting points.
   d. low boiling points and low melting points.

_____ **7.** Intermolecular forces include
   a. dipole-dipole forces only.
   b. London forces only.
   c. neither dipole-dipole nor London forces.
   d. both dipole-dipole and London forces.

_____ **8.** Dipole-dipole forces are especially significant when molecules are
   a. nonpolar.
   b. highly polar.
   c. slightly polar.
   d. ionic.

_____ **9.** The main reason that water contains especially strong hydrogen bonds is that
   a. hydrogen and oxygen differ greatly in electronegativity.
   b. the water molecule is linear.
   c. its boiling point is relatively high.
   d. its melting point is relatively low.

_____ **10.** The type of force between nonpolar molecules is a(n)
   a. covalent bond.
   b. hydrogen bond.
   c. ionic bond.
   d. London dispersion force.

_____ **11.** One mole of benzene vapor is cooled to its boiling point. The amount of energy released by the benzene as it completely condenses is its molar
   a. enthalpy of fusion.
   b. enthalpy of vaporization.
   c. entropy of fusion.
   d. entropy of vaporization.

_____ **12.** A spontaneous change is favored by an enthalpy change that is
   a. negative.
   b. positive.
   c. equal to $T\Delta S$.
   d. equal to the change in temperature.

_____ **13.** If the temperature is 100.0 K, $\Delta H = 2000$ J/mol, and $\Delta S = 22$ J/mol·K, a change in phase will
   a. be spontaneous.
   b. not be spontaneous.
   c. depend on other conditions.
   d. remain at equilibrium.

Chapter Test *continued*

_____**14.** If $\Delta S_{fus} = 30$ J/mol·K and $\Delta H_{fus} = 6000$ J/mol for a certain substance, its melting point is
  a. 0.005 K.
  b. 20 K.
  c. 200 K.
  d. 180 000 K.

_____**15.** Pressure changes affect the entropy of
  a. a solid.
  b. a liquid.
  c. a gas.
  d. a solid, a liquid, and a gas.

_____**16.** A two-phase system always consists of
  a. two distinct materials that each have the same composition and properties throughout.
  b. a solid and a liquid.
  c. the same material in two different states.
  d. two different pure substances.

_____**17.** At its boiling point in a closed system, water exists
  a. as a liquid only.
  b. as a gas only.
  c. with liquid and gas in equilibrium.
  d. with steam dissolved in the liquid.

_____**18.** A sample of carbon dioxide gas is in equilibrium with solid dry ice. If the temperature of the system increases, the
  a. vapor pressure of the gas increases.
  b. vapor pressure of the gas decreases.
  c. dry ice melts.
  d. carbon dioxide gas condenses.

_____**19.** A phase diagram relates the state of matter, pressure, and
  a. temperature.
  b. volume.
  c. mass.
  d. weight.

_____**20.** On a phase diagram, the point at which all equilibrium lines join is the
  a. melting point.
  b. boiling point.
  c. critical point.
  d. triple point.

Answer the following questions in the spaces provided.

21. Explain how cohesion relates to both surface tension and capillary action.

_____

_____

_____

_____

_____

_____

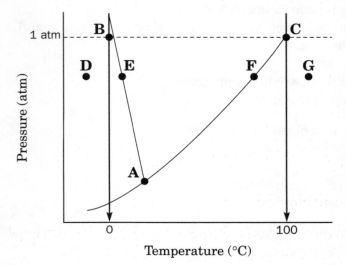

Figure 1

22. Explain what points B and C represent on **Figure 1**.

_____

_____

_____

23. In what phase is water at point D? at point G?

_____

_____

_____

24. Explain the locations of points E and F. What will happen to the phases of points E and F if the temperature is lowered?

_____

_____

_____

Answer the following problem in the space provided. Show all calculations.

25. For gold, $\Delta S_{fus} = 9.27$ J/mol·K and $\Delta H_{fus} = 12.4$ kJ/mol. What is the melting point of gold?

Quick Lab

Wetting a Surface

OBJECTIVES

Test the wetting ability of water on several surfaces.

MATERIALS

- detergent, diluted with water
- glass plate
- plastic plate
- steel plate
- toothpick
- water

Always wear safety goggles, gloves, and a lab apron to protect your eyes and clothing. If you get a chemical in your eyes, immediately flush the chemical out at the eyewash station while calling to your teacher. Know the location of the emergency lab shower and eyewash station and the procedures for using them.

Procedure

1. Wash plastic, steel and glass plates well by using dilute detergent and rinse them completely. Do not touch the clean surfaces.

2. Using a toothpick, put a small drop of water on each plate. Observe the shape of the drops from the side.

Analysis

1. On which surface does the water spread the most?

_____

2. On which surface does the water spread the least?

_____

3. What can you conclude about the adhesion of water for plastic, steel and glass?

_____

_____

4. Explain your observations in terms of wetting.

_____

_____

Viscosity of Liquids

SITUATION

You have been contacted by an automotive service shop that received a shipment of bulk containers of motor oil. The containers had been shipped by freight train, but several boxcars had leaky roofs. As a result, the labels peeled off the cans. Before the shop uses this oil in cars, the service technicians must match up the cans with the types of oil that were listed on the shipping invoice, based on the viscosity and the SAE rating of the oils.

BACKGROUND

Viscosity is the measurement of a liquid's resistance to flow. Several factors contribute to viscosity. Liquids with high intermolecular forces tend to be very viscous. For example, glycerol has a high viscosity because of its tendency to form many hydrogen bonds. For other molecules, such as oils, the longer the chain length of the molecule, the more viscous they are. The longer chains not only provide greater surface area for intermolecular attractions, but also can be intertwined more easily. For example, gasoline, which contains molecules that are chains of three to eight carbon atoms, is much less viscous than grease, which usually contains molecules with about 20 to 25 carbon atoms.

The Society of Automotive Engineers rates lubricating oils according to their comparative viscosities. These numerical values, called *SAE ratings*, range from SAE-10 (low viscosity) to SAE-60 (high viscosity) for oils typically used in combustion engines such as those in automobiles and trucks. The ratings are achieved with an instrument called a *viscosimeter*, which has a small capillary tube opening. The amount of time for a specific amount of motor oil to flow through the opening is a measure of viscosity. The less viscous oils flow through in a shorter time than the more viscous oils do.

PROBLEM

To match the correct oil sample to its SAE rating, you will need to do the following.

- Make your own viscosimeter from a pipet.
- Measure the relative viscosities of several oils by timing the oil as it flows through your viscosimeter.
- Measure mass and volume of each oil to calculate density.
- From the measurements, infer which labels belong on the containers of oil.

Name _____ Class _____ Date _____

OBJECTIVES

Demonstrate proficiency in comparing the viscosity of various liquids under identical test conditions.

Construct a small viscosimeter.

Measure flow time of various single-weight oils.

Measure the mass and volume of the oils to calculate density.

Calculate the relative viscosity of the oils.

Graph experimental data.

Compare viscosities and densities to determine the SAE rating of each oil.

MATERIALS

- beakers, 400 mL (2)
- beakers, 50 mL (7)
- distilled water
- gloves
- graduated cylinder, 10 mL
- ice
- lab apron
- oil samples, 10 mL (6)
- pin, straight
- pipets, thin-stem (7)
- ruler, metric
- safety goggles
- stopwatch or clock with second hand
- test-tube holder
- test-tube rack
- test tubes, small (7)
- wax pencil

Bunsen burner option
- Bunsen burner and related equipment
- ring stand and ring
- wire gauze with ceramic center

Hot plate option
- hot plate

Thermometer option
- thermometer, nonmercury
- thermometer clamp

Probe option
- thermistor probe

Always wear safety goggles, gloves, and a lab apron to protect your eyes and clothing. If you get a chemical in your eyes, immediately flush the chemical out at the eyewash station while calling to your teacher. Know the location of the emergency lab shower and eyewash station and the procedures for using them.

Do not touch any chemicals. If you get a chemical on your skin or clothing, wash the chemical off at the sink while calling to your teacher. Make sure you carefully read the labels and follow the precautions on all containers of chemicals that you use. If there are no precautions stated on the label, ask your teacher what precautions to follow. Do not taste any chemicals or items used in the laboratory. Never return leftovers to their original container; take only small amounts to avoid wasting supplies.

Name _____ Class _____ Date _____

Viscosity of Liquids *continued*

 Do not heat glassware that is broken, chipped, or cracked. Use tongs or a hot mitt to handle heated glassware and other equipment because hot glassware does not always look hot.

When using a Bunsen burner, confine long hair and loose clothing. If your clothing catches on fire, WALK to the emergency lab shower and use it to put out the fire. Because the oil tested in this lab is flammable, it should never be heated directly over a flame. Instead, use a hot-water bath, and never heat it above 60°C.

When heating a substance in a test tube, the mouth of the test tube should point away from where you and others are standing. Watch the test tube at all times to prevent the contents from boiling over.

Pins are sharp; use with care to avoid cutting yourself or others.

Procedure
PART 1–PREPARATION

1. Put on safety goggles, gloves, and a lab apron.

2. With a wax pencil, label each 50 mL beaker-test tube-pipet set with the name of one oil sample (*A, B, C, D, E,* or *F*). Label an additional set H_2O.

3. Place two marks 2.0 cm apart on the side of the bulb of the pipet, as shown in **Figure 1**. The top mark will be the starting point and the lower mark will be the endpoint.

4. Carefully make a small hole in the top of the bulb of each pipet with the pin, as shown in **Figure 1**. Be sure the hole is well above the marks you made on the side of the pipet bulb. Make the hole the same size for each pipet by putting in the pin the same way for each one. You will control the flow of oil with your finger and this hole.

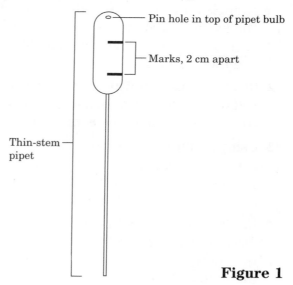

Figure 1

PART 2–TECHNIQUE

5. Measure the masses of all seven 50 mL beakers. Record them in your data table.

6. Pour about 5.0 mL of distilled water into the graduated cylinder. Measure and record the volume to the nearest 0.1 mL, and pour it into the H_2O beaker.

7. Measure and record the mass of the H_2O beaker with water in your data table.

Viscosity of Liquids *continued*

8. Squeeze the H_2O pipet bulb and fill the pipet with distilled water to above the top line. After it is full, place your finger over the pin hole. Place the pipet over the H_2O beaker, lift your finger off the hole, and allow the liquid to flow into the beaker until the meniscus is even with the top line on the pipet bulb. Cover the hole promptly when the water reaches this point. Several practice trials may be necessary.

9. One member of the lab group should hold the pipet with a finger over the pinhole, and the other should use a clock with a second hand or a stopwatch to record precise time intervals. Hold the pipet over the H_2O beaker. When the timer is ready, remove your finger from the pinhole, and allow the liquid to flow into the beaker until it reaches the bottom line on the pipet bulb. Record the time elapsed to the nearest 0.1 s in your data table in the section for room temperature. (If you do not have a stopwatch, measure the time elapsed to the nearest 0.5 s.) It may take several practice trials to master the technique.

10. Repeat **steps 6–9** with each oil, using the appropriately labeled pipets and beakers. You should perform several trials for each oil and for water to obtain consistent results. Clean the graduated cylinder after the last trial for each oil.

11. Using one of the 400 mL beakers, make an ice bath. Fill the test tubes to within 1.0 cm of the top with the appropriate oil or distilled water. Cool the samples for 5–8 min so that they are at a temperature between 0°C and 10°C. The key is that all of the samples must be at the same temperature. Measure the temperature of the water sample to the nearest 0.1°C with a thermometer or a thermistor probe and record it below your data table.

12. Repeat **steps 8–9** with each of the cooled samples. Be sure to use the pipets and 50 mL beakers designated for each oil or distilled water. Record the volume, mass, and time elapsed for each trial in your data table.

13. Using a Bunsen burner or a hot plate and the second 400 mL beaker, prepare a warm-water bath with a temperature between 35°C and 45°C. If you measure the temperature with a thermometer, use a thermometer clamp attached to a ring stand to hold the thermometer in the water.

14. Refill the test tubes to within 1.0 cm of the top with the appropriate oil or distilled water. Place these test tubes into the warm-water bath, and allow the oil and water to warm. Measure the temperature of the water sample with a thermometer or a thermistor probe when you remove the samples and record it below **Table 1.**

15. Repeat **steps 8–9** with the warm samples. Record the volume, mass, and time elapsed for each trial in your data table.

16. Your instructor will have set out twelve disposal containers; six for the six types of oil and six for the pipets. **Do not pour oil down the sink. Do not put the oil or oily pipets in the trash can.** The distilled water may be poured down the sink. The test tubes should be washed with a mild detergent and rinsed. Always wash your hands thoroughly after cleaning up the area and equipment.

| Viscosity of Liquids *continued*

TABLE 1 FLOW TIMES OF THE OILS

Sample	Beaker mass (g)	Total mass (g)	Volume (mL)	Trial 1— cool (s)	Trial 2— cool (s)	Trial 3— cool (s)
A						
B						
C						
D						
E						
F						
H_2O						

Sample	Trial 1— room temp. (s)	Trial 2— room temp. (s)	Trial 3— room temp. (s)	Trial 1— warm (s)	Trial 2— warm (s)	Trial 3— warm (s)
A						
B						
C						
D						
E						
F						
H_2O						

cool temperature: _____

room temperature: _____

warm temperature: _____

Analysis

1. **Organizing Data** Determine the density of each sample.

A: _____   E: _____

B: _____   F: _____

C: _____ H_2O: _____

D: _____

Name _____ Class _____ Date _____

Viscosity of Liquids *continued*

2. **Organizing Data** Find the average flow time for each sample at each temperature.

Sample	Avg. time (s) cool	Avg. time (s) room temp.	Avg. time (s) warm
A			
B			
C			
D			
E			
F			
H_2O			

3. **Analyzing Information** Calculate the relative viscosity of your samples at room temperature by applying the following formula. The values for the absolute viscosity of water are in units of centipoises (cp). A centipoise is equal to 0.01 g/cm·s.

$$\text{relative viscosity}_{oil} = \frac{\text{density}_{oil} \times \text{time elapsed}_{oil} \times \text{viscosity}_{H_2O}}{\text{density}_{H_2O} \times \text{time elapsed}_{H_2O}}$$

Temperature (°C)	Absolute Viscosity for H_2O (cp)
18	1.053
20	1.002
22	0.955
24	0.911
25	0.890
26	0.870
28	0.833

relative viscosity$_A$: _____

relative viscosity$_B$: _____

relative viscosity$_C$: _____

relative viscosity$_D$: _____

relative viscosity$_E$: _____

relative viscosity$_F$: _____

Viscosity of Liquids *continued*

Conclusions

4. **Inferring Conclusions** According to the invoice, the service station was supposed to receive equal amounts of SAE-10, SAE-20, SAE-30, SAE-40, SAE-50, and SAE-60 oil. Given that the oils with the lower SAE ratings have lower relative viscosities, infer which oil samples correspond to the SAE ratings indicated.

_____

_____

_____

Letter	Rel. viscosity (cp)	Oil type
A		
B		
C		
D		
E		
F		

5. **Organizing Information** Prepare a graph with flow time at room temperature on the *y*-axis and SAE rating on the *x*-axis.

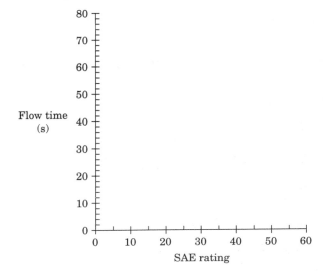

Viscosity of Liquids *continued*

6. Organizing Information Prepare a graph with density on the *y*-axis and SAE rating on the *x*-axis.

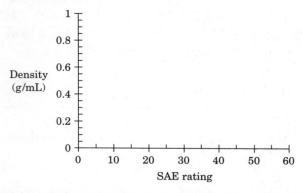

7. Organizing Information Prepare a graph with viscosity at room temperature on the *y*-axis and SAE rating on the *x*-axis.

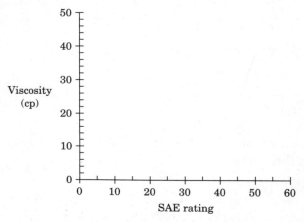

8. Organizing Information Prepare a graph with viscosity at room temperature on the *y*-axis and density on the *x*-axis.

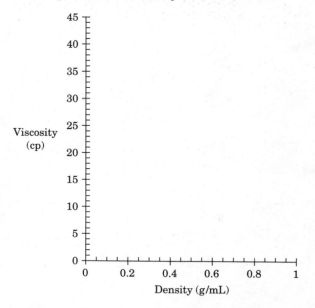

Viscosity of Liquids *continued*

9. Inferring Conclusions How does temperature affect the viscosity of each sample?

_____

_____

_____

_____

10. Interpreting Graphics Is there a relationship between density and viscosity?

_____

_____

_____

_____

11. Interpreting Graphics What is the relationship between SAE rating and viscosity?

_____

_____

_____

_____

12. Interpreting Graphics What is the relationship between viscosity and flow time?

_____

_____

_____

_____

Extensions

1. Predicting Outcomes Estimate what flow times you would measure at each temperature if you repeated the tests in this lab with SAE-35 oil.

_____

_____

_____

_____

Viscosity of Liquids *continued*

2. Relating Ideas Malcolm is trying to get the last of the pancake syrup out of a bottle. What can he do to make the syrup come out of the bottle faster? Explain how your plan will take advantage of viscosity.

_____

_____

_____

3. Research and Communication Contact a manufacturer of lubrication products such as Valvoline or Pennzoil, and write a short paper on the development and properties of the oils used in this investigation.

_____

_____

_____

_____

Skills Practice Lab

Constructing a Heating/Cooling Curve

SITUATION

Sodium thiosulfate pentahydrate, $Na_2S_2O_3 \cdot 5H_2O$, is produced by a local manufacturing firm and sold nationwide to photography shops, paper processing plants, and textile manufacturers. Purity is one condition of customer satisfaction, so samples of $Na_2S_2O_3 \cdot 5H_2O$ are taken periodically from the production line and tested for purity by an outside testing facility. Your company has been tentatively chosen because your proposal was the only one based on melting and freezing points rather than the more expensive titrations with iodine. To make the contract final, you must convince the manufacturing firm that you can establish accurate standards for comparison.

BACKGROUND

As energy flows from a liquid, its temperature drops. The entropy, or random ordering of its particles, also decreases until a specific ordering of the particles results in a phase change to a solid. If energy is being released or absorbed by a substance remaining at the same temperature, this is evidence that a dramatic change in entropy, such as a phase change, is occurring. Because all of the particles of a pure substance are identical, they all freeze at the same temperature, and the temperature will not change until the phase change is complete. If a substance is impure, the impurities will not lose energy in the same way that the rest of the particles do. Therefore, the freezing point will be somewhat lower, and there will be a range of temperatures instead of a single temperature.

PROBLEM

To evaluate the samples, you will need a heating/cooling curve for pure $Na_2S_2O_3 \cdot 5H_2O$ that you can use as a standard. To create and use this curve, you must do the following.

- Obtain a measured amount of pure $Na_2S_2O_3 \cdot 5H_2O$.

- Melt and freeze the sample, periodically recording the time and temperature.

- Graph the data to determine the melting and freezing points of pure $Na_2S_2O_3 \cdot 5H_2O$.

- Interpret the changes in energy and entropy involved in these phase changes.

- Verify the observed melting point against the accepted melting point found in reference data from two different sources.

- Use the graph to qualitatively determine whether there are impurities in a sample of $Na_2S_2O_3 \cdot 5H_2O$.

Constructing a Heating/Cooling Curve *continued*

OBJECTIVES

Observe the temperature and phase changes of a pure substance.

Measure the time needed for the melting and freezing of a specified amount of substance.

Graph experimental data and determine the melting and freezing points of a pure substance.

Analyze the graph for the relationship between melting point and freezing point.

Identify the relationship between temperature and phase change for a substance.

Infer the relationship between energy and phase changes.

Recognize the effect of an impurity on the melting point of a substance.

Analyze the relationship between energy, entropy, and temperature.

MATERIALS

- balance, centigram
- beaker tongs
- beakers, 600 mL (3)
- chemical reference books
- forceps
- gloves
- graph paper
- hot mitt
- ice
- lab apron
- $Na_2S_2O_3 \cdot 5H_2O$
- plastic washtub
- ring clamps (3)
- ring stands (2)
- ruler
- safety goggles

- stopwatch or clock with a second hand
- test-tube clamp
- test tube, Pyrex, medium
- thermometer clamp
- wire gauze with ceramic center (2)
- wire stirrer

Bunsen burner option
- Bunsen burner
- gas tubing
- striker

Hot plate option
- hot plate

Probe option
- thermistor probes (2)

Thermometer option
- thermometers, nonmercury (2)

Always wear safety goggles, gloves, and a lab apron to protect your eyes and clothing. If you get a chemical in your eyes, immediately flush the chemical out at the eyewash station while calling to your teacher. Know the location of the emergency lab shower and eyewash station and the procedures for using them.

Constructing a Heating/Cooling Curve *continued*

Do not touch any chemicals. If you get a chemical on your skin or clothing, wash the chemical off at the sink while calling to your teacher. Make sure you carefully read the labels and follow the precautions on all containers of chemicals that you use. If there are no precautions stated on the label, ask your teacher what precautions to follow. Do not taste any chemicals or items used in the laboratory. Never return leftovers to their original container; take only small amounts to avoid wasting supplies.

Do not heat glassware that is broken, chipped, or cracked. Use tongs or a hot mitt to handle heated glassware and other equipment because hot glassware does not always look hot.

When using a Bunsen burner, confine long hair and loose clothing. If your clothing catches on fire, WALK to the emergency lab shower and use it to put out the fire.

When heating a substance in a test tube, the mouth of the test tube should point away from where you and others are standing. Watch the test tube at all times to prevent the contents from boiling over.

Procedure
PART 1–PREPARATION

1. Put on safety goggles, gloves, and a lab apron.

2. Fill two 600 mL beakers three-fourths full of tap water.

3. Heat water for a hot-water bath. If you are using a Bunsen burner, attach to a ring stand a ring clamp large enough to hold a 600 mL beaker. Adjust the height of the ring until it is 10 cm above the burner. Cover the ring with wire gauze. Set one 600 mL beaker of water on the gauze. If you are using a hot plate, rest the beaker of water directly on the hot plate.

4. Monitor the temperature of the water with a thermometer or a thermistor probe. Complete **steps 5–8** while the water is heating.

5. Cool the water for a cold-water bath. Fill a small plastic washtub with ice. Form a hole in the ice that is large enough for the second 600 mL beaker. Insert the beaker and pack the ice around it up to the level of the water in the beaker.

6. Bend the piece of wire into the shape of a stirrer, as shown in **Figure 1.** One loop should be narrow enough to fit into the test tube, yet wide enough to easily fit around the thermometer without touching it.

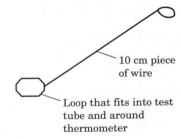

10 cm piece of wire

Loop that fits into test tube and around thermometer

Figure 1

Name _____ Class _____ Date _____

Constructing a Heating/Cooling Curve *continued*

7. Prepare the sample. Assemble the test tube, thermometer, and stirrer, as shown in **Figure 2.** Attach the entire assembly to a second ring stand. Then, add enough $Na_2S_2O_3 \cdot 5H_2O$ crystals so that the test tube is about one-quarter full and the thermometer bulb is well under the surface of the crystals as shown in **Figure 3.**

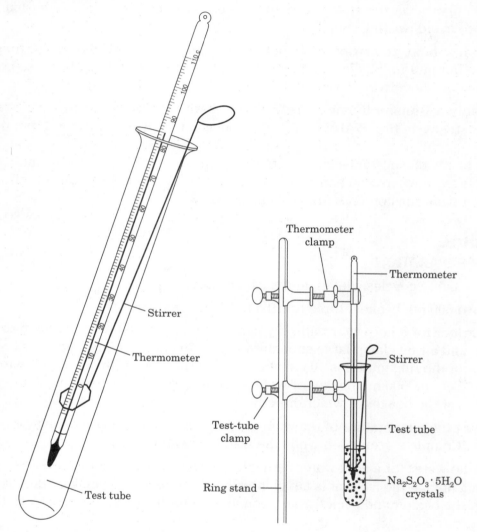

Figure 2

Figure 3

8. Set up the container for the hot-water bath as shown in **Figure 4.** Attach two ring clamps, one above the other, to the second ring stand beneath the test-tube assembly. Place a wire gauze with ceramic center on the lower ring. Set a third 600 mL beaker, which should be empty, on the gauze and raise the beaker toward the test-tube assembly until it surrounds nearly one-half of the tube's length. The beaker will pass through the ring clamp without gauze, and the test tube should not touch the bottom or sides of the beaker, as shown in **Figure 4** on the next page. The top clamp keeps the beaker from tipping when the beaker is filled with the hot water.

Constructing a Heating/Cooling Curve *continued*

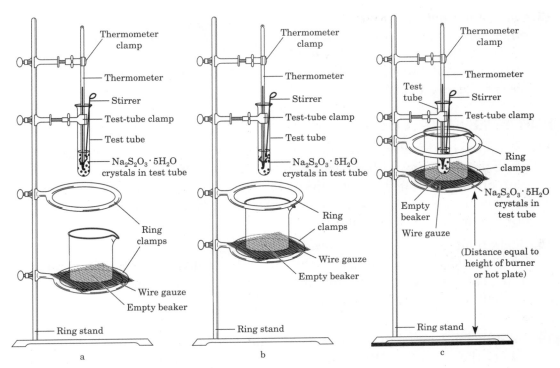

Figure 4

PART 2–MELTING A SOLID: QUICK TEST

9. Check the temperature of the water for the hot-water bath. When it is 85°C, turn off the burner or hot plate. If the temperature is already greater than 85°C, shut off the burner or hot plate, and add a few pieces of ice to bring the temperature down to 85°C. Then, using beaker tongs, remove the beaker of hot water from the burner. Using tongs or a hot mitt carefully pour the water into the empty beaker until the water level is well above the level of the solid inside the test tube. Set the empty beaker on the counter. You will use it again in step **20.**

10. Begin timing. The second the water is poured, one member of the lab group should begin timing, while the other reads the initial temperatures of the bath with one thermometer and sample with the other thermometer.

11. Occasionally stir the melting solid by gently moving the stirrer up and down. Be careful not to break the thermometer bulb. Monitor the temperature of the $Na_2S_2O_3 \cdot 5H_2O$ and the hot-water bath with separate thermometers or probes.

12. When the temperature of the liquid $Na_2S_2O_3 \cdot 5H_2O$ is approximately the same as that of the hot-water bath, stop timing. Note the final temperature of the liquid $Na_2S_2O_3 \cdot 5H_2O$ and the elapsed time. This temperature is the approximate melting point of your sample. Knowing this value can help you make the careful observations necessary to determine a more precise value.

Constructing a Heating/Cooling Curve *continued*

13. Using a hot mitt, hold the beaker of hot water with one hand while using the other hand to gently loosen only the lower ring clamp enough so that the beaker of hot water can be lowered and removed. Remove the beaker of hot water, set it on the gauze above the burner, and let it reheat to 65°C while you perform **steps 14–20.**

PART 3—FREEZING A LIQUID

14. Set up the cold-water bath. Remove the beaker of cold water from the ice and place it on the ring with the gauze, well below the test tube. Steady the beaker with one hand while raising it until the level of the cold water is well above the level of the liquid inside the test tube. The test tube should not touch the bottom or sides of the beaker.

15. Begin timing. The second that the cold water is in place, one member of the lab group should begin timing, while the other reads the initial temperatures of the sample and the bath. Record the initial time and temperatures in the left half of **Table 1.** The starting temperature of the liquid should be near 80°C.

16. Monitor the cooling process. Measure and record the time and the temperature of the $Na_2S_2O_3 \cdot 5H_2O$ every 15 s in the left half of **Table 1.** Also record observations about the substance's appearance and other properties in the *Observations of cooling* column in **Table 1.** When the temperature reaches 50°C, use forceps to add one or two seed crystals of $Na_2S_2O_3 \cdot 5H_2O$ to the test tube.

17. Continue taking temperature readings every 15 s, stirring continuously, until a constant temperature is attained. (A temperature is constant if it is recorded at four consecutive 15 s intervals.) **Do not try to move the thermometer, thermistor probe, or stirrer when solidification occurs.**

18. Finish timing. Continue taking readings until the temperature of the solid differs from the temperature of the cold-water bath by 5°C.

19. Remove the cold-water bath. Grasp the beaker with one hand, carefully loosen its supporting ring clamp with the other hand, and lower the beaker of cold water away from the test tube. Remove the beaker from the ring and set it on the counter.

PART 3—MELTING A SOLID

20. Set up the container for the hot-water bath. Place the empty beaker from **step 9** on the ring and wire gauze. Steady the beaker as you raise it to surround the test tube as you did in **step 8,** but this time allow room for the Bunsen burner to be placed under the beaker.

Constructing a Heating/Cooling Curve *continued*

21. Fill the hot-water bath. Use the second thermometer or thermistor probe to check the temperature of the water for the hot-water bath. When it is 65°C, turn off the burner or hot plate. If the temperature is greater than 65°C, add a few pieces of ice to lower the temperature. Using tongs or a hot mitt, carefully pour the hot water into the empty beaker until the water level is well above the level of the solid inside the test tube. Set the empty beaker on the counter.

22. Begin timing. The second that the water is poured, one member of the lab group should begin timing while another reads initial temperatures of the water bath and the solid $Na_2S_2O_3 \cdot 5H_2O$. Record the solid's temperature in the right half of the data table. The starting temperature of the solid should be below 35°C.

23. Maintain the bath's temperature. Move the burner or hot plate under the hot-water bath and continue heating the water in the bath. Adjust the position and size of the flame or the setting of the hot plate so that the temperature of the hot-water bath remains between 60°C and 65°C.

24. Monitor the warming process. Record the temperature of the sample every 15 s. Use the stirrer, when it becomes free of the solid, to gently stir the contents of the test tube. Also record observations about the substance's appearance and other properties in the *Observations of warming* column in **Table 1.**

25. Continue taking readings until the temperature of the $Na_2S_2O_3 \cdot 5H_2O$ differs from that of the hot-water bath by 5°C.

26. Record the final temperature and the time.

27. Turn off the burner or hot plate.

28. Remove the thermometer or thermistor probe from the liquid $Na_2S_2O_3 \cdot 5H_2O$ and rinse it. Pour the $Na_2S_2O_3 \cdot 5H_2O$ from the test tube into the disposal container designated by your teacher. If you used a Bunsen burner, check to see that the gas valve is completely turned off. Remember to wash your hands thoroughly after cleaning up the lab area and all equipment.

Constructing a Heating/Cooling Curve *continued*

TABLE 1 TIME AND TEMPERATURE DATA

Cooling Data			Warming Data		
Time (s)	Temp. (°C)	Observations of cooling	Time (s)	Temp. (°C)	Observation of warming
0			0		
15			15		
30			30		
45			45		
60			60		
75			75		
90			90		
105			105		
120			120		
135			135		
150			150		
165			165		
180			180		
195			195		
210			210		
225			225		
240			240		
255			255		
270			270		
285			285		
300			300		
315			315		
330					
345					
360					
375					
390					

Constructing a Heating/Cooling Curve *continued*

Analysis

1. Organizing Data Plot both the heating and cooling data on the same graph. Place time on the *x*-axis and temperature on the *y*-axis.

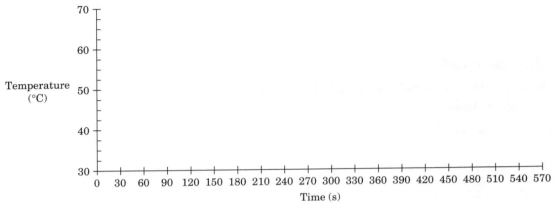

2. Interpreting Graphics Describe and compare the shape of the cooling curve with the shape of the heating curve.

_____

_____

_____

_____

_____

3. Interpreting Graphics Locate the freezing and melting temperatures on your graph. Compare them and comment on why they have different names.

_____

_____

_____

_____

_____

4. Evaluating Methods One purpose of the quick test for melting point is summarized in **step 12**. State this purpose and explain how it prepares you for **steps 17 and 24.**

_____

_____

_____

Constructing a Heating/Cooling Curve *continued*

5. Evaluating Data Compare your melting point with that found in references. What is your percent error?

Conclusions

6. Analyzing Information As the liquid cools, what is happening to the kinetic energy and the entropy of the following?

a. $Na_2S_2O_3 \cdot 5H_2O$

_____

_____

b. the water bath

_____

_____

7. Analyzing Information What happened to the temperature of the sample from the time that freezing began until freezing was complete? Did the entropy of the sample increase, decrease, or stay the same?

_____

_____

_____

8. Predicting Outcomes How would the quantity of the sample affect the time needed for the melting point test?

_____

_____

_____

9. Predicting Outcomes Would the quantity of the sample used to determine the melting point affect its outcome? (Hint: is melting point an extensive or intensive property?)

_____

_____

_____

Constructing a Heating/Cooling Curve *continued*

10. Interpreting Graphics

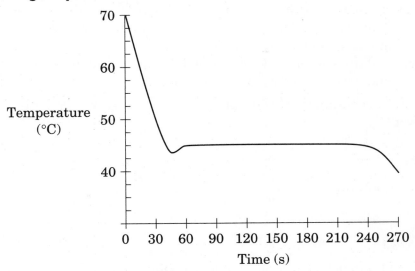

Examine the graph above, and compare it to your cooling curve. Would this sample of sodium thiosulfate pentahydrate be considered pure or impure? Sketch a line on the graph that represents your cooling data. If the curves are not identical, estimate the difference in melting points.

_____

_____

_____

Extensions

1. **Interpreting Graphics** Refer to the heating and cooling curves you plotted. For each portion of the curve, describe what happens to the energy and entropy of the substance.

_____

_____

_____

_____

_____

_____

_____

Name _____ Class _____ Date _____

Constructing a Heating/Cooling Curve *continued*

2. **Applying Ideas** In northern climates, freezing rain is a driving hazard. When this occurs, warm air from a defroster is blown against the windshield of an automobile in order to restore visibility. It would be convenient to have a system that automatically turned the defroster blower system on and off as needed. A thermostat embedded in the windshield to detect outside temperature could be used to perform this function.

   a. At what temperature should the thermostat be set to turn on the hot-air blower?

   _____

   _____

   _____

   _____

   b. At what temperature should the thermostat be set to turn off the hot-air blower?

   _____

   _____

   _____

   _____

3. **Analyzing Methods** Will crystallization take place if no seed crystal is added? Why or why not?

   _____

   _____

   _____

   _____

4. **Designing Experiments** Explain the purpose of a water bath. Why wasn't distilled water necessary?

   _____

   _____

   _____

   _____

Constructing a Heating/Cooling Curve *continued*

5. Organizing Ideas Which of the following word equations best represents the changes of phase taking place in the situations described in items **1** to **9** below? Place your answer in the space to the left of the numbered items.

    a. solid + energy ⟶ liquid

    b. liquid ⟶ solid + energy

    c. solid + energy ⟶ vapor

    d. vapor ⟶ solid + energy

_____ **1.** ice melting at 0°C

_____ **2.** water freezing at 0°C

_____ **3.** a mixture of ice and water whose relative amounts remain unchanged

_____ **4.** a particle escaping from a solid and becoming a vapor particle

_____ **5.** solids, like camphor and naphthalene, subliming

_____ **6.** snow melting

_____ **7.** snow forming

_____ **8.** dry ice subliming

_____ **9.** dry ice forming

Skills Practice Lab

Evaporation and
Intermolecular Attractions

PROBEWARE LAB

You are an organic chemist working for a company that manufactures various types of ink. You have been asked to create a calligrapher's ink that dries quickly at room temperature. The company feels that such a product would be a big hit since faster-drying ink would cause less distortion to the paper on which it is used. Ink consists of two components: a pigment, or coloring agent, and a solvent. The pigment is what gives the ink its color. The solvent is the chemical in which the pigment is dissolved. Your job is to select a solvent that will evaporate quickly.

To determine the best solvent to use, you will test two types of organic compounds—alkanes and alcohols. To establish how and why these substances evaporate, you will test four alcohols and two alkanes. From your results, you will be able to predict how other alcohols and alkanes will evaporate.

The two alkanes you will test are pentane, C_5H_{12}, and hexane, C_6H_{14}. Alkanes contain only carbon and hydrogen atoms, while alcohols also contain the –OH functional group. In this experiment, two of the alcohols you will test are methanol, CH_3OH, and ethanol, C_2H_5OH. To better understand why these substances evaporate, you will examine the molecular structure of each for the presence and relative strength of hydrogen bonding and London dispersion forces.

The process of evaporation requires energy to overcome the intermolecular forces of attraction. For example, when you perspire on a hot day, the water molecules in your perspiration absorb heat from your body and evaporate. The result is a lowering of your skin temperature known as evaporative cooling.

In this experiment, temperature probes will be placed into small containers of your test substances. When the probes are removed, the liquid on the temperature probes will evaporate. The temperature probes will monitor the temperature change. Using your data, you will determine the temperature change, ΔT, for each substance and relate that information to the substance's molecular structure and presence of intermolecular forces.

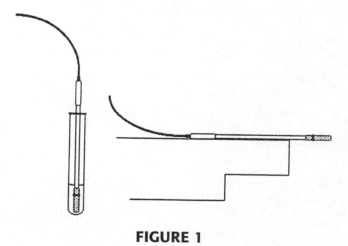

FIGURE 1

| Evaporation and Intermolecular Attractions *continued*

OBJECTIVES

- **Measure** temperature changes.
- **Calculate** changes in temperature.
- **Relate** temperature changes to molecular bonding.
- **Predict** temperature changes for various liquids.

MATERIALS

- 1-butanol
- ethanol (ethyl alcohol)
- *n*-hexane
- *n*-pentane
- 1-propanol

- methanol (methyl alcohol)
- filter paper pieces, 2.5 cm × 2.5 cm (6)
- masking tape
- rubber bands, small (2)

EQUIPMENT

- LabPro or CBL2 interface
- TI graphing calculator
- Vernier temperature probes (2)

SAFETY

- Wear safety goggles when working around chemicals, acids, bases, flames, or heating devices. Contents under pressure may become projectiles and cause serious injury.

- If any substance gets in your eyes, notify your instructor immediately and flush your eyes with running water for at least 15 minutes.

- Use flammable liquids only in small amounts.

- When working with flammable liquids, be sure that no one else in the lab is using a lit Bunsen burner or plans to use one. Make sure there are no other heat sources present.

- Secure loose clothing, and remove dangling jewelry. Do not wear open-toed shoes or sandals in the lab.

- Wear an apron or lab coat to protect your clothing when working with chemicals.

- Never return unused chemicals to the original container; follow instructions for proper disposal.

- Always use caution when working with chemicals.

- Never mix chemicals unless specifically directed to do so.

- Never taste, touch, or smell chemicals unless specifically directed to do so.

Pre-lab Procedure

Before doing the experiment, complete the pre-lab table. The name and formula are given for each compound. Draw a structural formula for a molecule of each compound. Then determine the molecular weight of each of the molecules. Dispersion forces exist between any two molecules and generally increase as the molecular weight of the molecule increases. Next, examine each molecule for the presence of hydrogen bonding. Before hydrogen bonding can occur, a hydrogen atom must be bonded directly to an N, O, or F atom within the molecule. Tell whether or not each molecule has hydrogen-bonding capability.

Substance	Formula	Structural formulas	Molecular weight	Hydrogen bond (yes or no)
Ethanol	C_2H_5OH			
1-propanol	C_3H_7OH			
1-butanol	C_4H_9OH			
n-pentane	C_5H_{12}			
Methanol	CH_3OH			
n-hexane	C_6H_{14}			

Procedure
EQUIPMENT PREPARATION

1. Obtain and wear goggles! **CAUTION:** *The compounds used in this experiment are flammable and poisonous. Avoid inhaling their vapors. Avoid their contact with your skin or clothing. Be sure there are no open flames in the lab during this experiment. Notify your teacher immediately if an accident occurs.*

2. Plug temperature probe 1 into Channel 1 and temperature probe 2 into Channel 2 of the LabPro or CBL 2 interface. Use the link cable to connect the TI graphing calculator to the interface. Firmly press in the cable ends.

3. Turn on the calculator, and start the DATAMATE program. Press `CLEAR` to reset the program.

4. Set up the calculator and interface for two temperature probes.

   a. Select SETUP from the main screen.

   b. If the calculator displays two temperature probes, one in CH 1 and another in CH 2, proceed directly to Step 5. If it does not, continue with this step to set up your sensor manually.

   c. Press `ENTER` to select CH 1.

   d. Select TEMPERATURE from the SELECT SENSOR menu.

   e. Select the temperature probe you are using (in °C) from the TEMPERATURE menu.

Evaporation and Intermolecular Attractions *continued*

    f. Press ▼ once, then press ENTER to select CH2.

    g. Select TEMPERATURE from the SELECT SENSOR menu.

    h. Select the temperature probe you are using (in °C) from the TEMPERATURE menu.

5. Set up the data-collection mode.

    a. To select MODE, use ▲ to move the cursor to MODE and press ENTER.

    b. Select TIME GRAPH from the SELECT MODE menu.

    c. Select CHANGE TIME SETTINGS from the TIME GRAPH SETTINGS menu.

    d. Enter "3" as the time between samples in seconds.

    e. Enter "80" as the number of samples. (The length of the data collection will be four minutes.)

    f. Select OK to return to the setup screen.

    g. Select OK again to return to the main screen.

6. Wrap probe 1 and probe 2 with square pieces of filter paper secured by small rubber bands as shown in **Figure 1.** Roll the filter paper around the probe tip in the shape of a cylinder. **Hint:** First slip the rubber band up on the probe, wrap the paper around the probe, and then finally slip the rubber band over the wrapped paper. The paper should be even with the probe end.

7. Stand probe 1 in the ethanol container and probe 2 in the 1-propanol container. Make sure the containers do not tip over.

8. Prepare two pieces of masking tape, each about 10 cm long, to be used to tape the probes in position during Step 9.

DATA COLLECTION

9. After the probes have been in the liquids for at least 30 seconds, select START to begin collecting temperature data. A live graph of temperature versus time for both Probe 1 and probe 2 is being plotted on the calculator screen. The live readings are displayed in the upper-right corner of the graph, the reading for probe 1 first, the reading for probe 2 below it. Monitor the temperature for 15 seconds to establish the initial temperature of each liquid. Then simultaneously remove the probes from the liquids, and tape them so the probe tips extend 5 cm over the edge of the table top as shown in **Figure 1.**

10. Data collection will stop after four minutes (or press the STO▶ key to stop *before* four minutes have elapsed). On the displayed graph of temperature versus time, each point for probe 1 is plotted with a dot, and each point for probe 2 with a box. As you move the cursor right or left, the time (X) and temperature (Y) values of each probe 1 data point are displayed below the graph. Based on your data, determine the maximum temperature, T_1, and minimum temperature, T_2. Record T_1 and T_2 for probe 1.

    Press ▼ to switch the cursor to the curve of temperature versus time for probe 2. Examine the data points along the curve. Record T_1 and T_2 for probe 2.

Evaporation and Intermolecular Attractions *continued*

11. For each liquid, subtract the minimum temperature from the maximum temperature to determine ΔT, the temperature change during evaporation.

12. Roll the rubber band up the probe shaft, and dispose of the filter paper as directed by your instructor.

13. Based on the ΔT values you obtained for these two substances, plus information in the pre-lab exercise, *predict* the ΔT value for 1-butanol. Compare its hydrogen-bonding capability and molecular weight with those of ethanol and 1-propanol. Record your predicted ΔT, and then explain how you arrived at this answer in the space provided. Do the same for *n*-pentane. It is not important that you predict the exact ΔT value; simply estimate a logical value that is higher, lower, or between the previous ΔT values.

14. Press ENTER to return to the main screen. Test your prediction in Step 13 by repeating Steps 6–12 using 1-butanol with probe 1 and *n*-pentane with probe 2.

15. Based on the ΔT values you have obtained for all four substances, plus information in the pre-lab exercise, predict the ΔT values for methanol and *n*-hexane. Compare the hydrogen-bonding capability and molecular weight of methanol and *n*-hexane with those of the previous four liquids. Record your predicted ΔT, and then explain how you arrived at this answer in the space provided.

16. Press ENTER to return to the main screen. Test your prediction in Step 15 by repeating Steps 6–12, using methanol with probe 1 and *n*-hexane with probe 2.

DATA TABLE

Substance	T_1 (°C)	T_2 (°C)	ΔT ($T_1 - T_2$) (°C)	Predicted ΔT(°C)	Explanation
Ethanol					
1-propanol					
1-butanol					
n-pentane					
Methanol					
n-hexane					

Evaporation and Intermolecular Attractions *continued*

Analysis

1. Analyzing data Which of the tested alcohols evaporated the fastest? Which alcohol had the largest ΔT value? What was the alcohol's molecular weight?

_____

_____

_____

2. Analyzing data Which of the alcohols tested evaporated the slowest? Which alcohol had the smallest ΔT value and molecular weight? _____

_____

_____

_____

3. Analyzing results The alcohol 1-butanol and alkane *n*-pentane have similar molecular weights, but their tests resulted in very different ΔT values. Based on the information in your pre-lab data table, explain the difference in the ΔT values of these substances. _____

_____

_____

_____

4. Analyzing information What types of intermolecular forces are evident in this experiment? _____

_____

_____

5. Analyzing data Make a graph of your data with the molecular weight of each substance on the *x*-axis and ΔT on the *y*-axis.

Conclusions

1. **Evaluating results** Alcohols with more than one –OH group are known as glycols. The substance ethylene glycol, CH_2OHCH_2OH, has a molecular weight of 62. Based on your data, would you expect it to have a larger or smaller ΔT than 1-propanol? Explain your answer using the results of this experiment.

_____

_____

_____

_____

_____

2. **Inferring conclusions** Of the substances tested in this experiment, which would work best as the solvent for the ink your company is developing?

_____

_____

Extensions

1. **Applying results** Using methanol and ethanol as solvent choices, test different types of mediums. Use the different medium choices in place of the filter paper to determine if the medium has any effect on the rate of evaporation.

Lesson Plan

Section: States and State Changes

Pacing

Regular Schedule	with lab(s): 4 days	without lab(s): 2 days
Block Schedule	with lab(s): 2 days	without lab(s): 1 day

Objectives

1. Relate the properties of a state to the energy content and particle arrangement of that state of matter.

2. Explain forces and energy changes involved in change of state.

National Science Education Standards Covered
UNIFYING CONCEPTS AND PROCESSES

UCP 1 Systems, order, and organization

UCP 3 Change, constancy, and measurement

UCP 5 Form and function

PHYSICAL SCIENCE—STRUCTURE AND PROPERTIES OF MATTER

PS 2e Solids, liquids, and gases differ in the distances and angles between molecules or atoms and therefore the energy that binds them together. In solids, the structure is nearly rigid; in liquids, molecules or atoms move around each other but do not move apart; and in gases, molecules or atoms move almost independently of each other and are mostly far apart.

> **KEY**
> **SE** = Student Edition
> **ATE** = Annotated Teacher Edition

Block 1 *(45 minutes)*
FOCUS *10 minutes*

❏ **Bellringer,** ATE (GENERAL). Ask students to discuss surface tension.

MOTIVATE *10 minutes*

❏ **Demonstration,** ATE (GENERAL). This demonstration illustrates how a magnetized needle will stay on top of water because of surface tension.

TEACH *25 minutes*

❏ **Using the Figure,** ATE (GENERAL). Compare the densities and arrangements of particles in Figure 2.

❏ **Transparency,** Mercury in Three States (GENERAL). This transparency master shows mercury in its gaseous, liquid, and solid states.

❏ **QuickLab, Wetting a Surface,** SE (GENERAL). Students observe the effects of detergent on surface tension. Five analysis questions follow the procedure.

❏ **Observation Lab, Viscosity of Liquids,** Chapter Resource File (GENERAL). Students observe temperature and phase changes and graph their data to determine the melting freezing points of a pur substance.

HOMEWORK

❏ **Reading Skill Builder,** ATE (BASIC). Have students list things that they already know about intermolecular forces and changes of state.

OTHER RESOURCES

❏ **Group Activity,** ATE (ADVANCED). This activity has students research and construct models of the seven different crystal systems shown in solids.

❏ **go.hrw.com**

❏ **www.scilinks.org**

Block 2 *(45 minutes)*
TEACH *30 minutes*

❏ **Transparency,** Changes of State. (GENERAL) This transparency master shows that most substances exist in three states and can change from state to state.

❏ **Demonstration,** ATE (GENERAL). This demonstration compares the meniscus in water and mercury.

❏ **Misconception Alert,** ATE (GENERAL). Use this feature to help students distinguish between melting and dissolving.

❏ **Using the Figure,** ATE (GENERAL). Have students use Figure 8 to discuss exothermic processes.

❏ **Misconception Alert,** ATE (GENERAL). Use this feature to help students distinguish between the physical change of vaporization and the chemical change of a gas being released.

CLOSE *15 minutes*

❏ **Reteaching,** ATE (BASIC). Students describe the changes of state involved in a day of many given weather changes.

❏ **Quiz,** ATE (GENERAL). This assignment has students answer questions about the concepts in this lesson.

❏ **Assessment Worksheet: Section Quiz** (GENERAL)

HOMEWORK

❑ **Skills Worksheet: Concept Review** (GENERAL) This worksheet reviews the main concepts and problem-solving skills of this section.

❑ **Homework,** ATE (BASIC). This assignment has students research a biological or geological cycle.

❑ **Skill Builder,** ATE (BASIC). Have students research the meaning of the word *volatile.*

❑ **Section Review,** SE (GENERAL). Assign items 1–15.

OTHER RESOURCES

❑ **Skill Builder,** ATE (ADVANCED). Have students write an explanation of why bubbles of steam are not observed rising to the surface of the ocean.

❑ **go.hrw.com**

❑ **www.scilinks.org**

Lesson Plan

Section: Intermolecular Forces

Pacing

Regular Schedule   **with lab(s):** 3½ days   **without lab(s):** 2 days
Block Schedule   **with lab(s):** 2 days   **without lab(s):** 1 day

Objectives

1. Contrast ionic and molecular substances in terms of their physical characteristics and the types of forces that govern their behavior.

2. Describe dipole-dipole forces.

3. Explain how a hydrogen bond is different from other dipole-dipole forces, and how it is responsible for many of water's properties.

4. Describe London dispersion forces, and relate their strength to other types of attractions.

National Science Education Standards Covered

UNIFYING CONCEPTS AND PROCESSES

UCP 1 Systems, order, and organization

UCP 3 Change, constancy, and measurement

UCP 5 Form and function

PHYSICAL SCIENCE—STRUCTURE AND PROPERTIES OF MATTER

PS 2d The physical properties of compounds reflect the nature of the interactions among their molecules. These interactions are determined by the structure of the molecule, including the constituent atoms and the distances and angles between them.

PS 2e Solids, liquids, and gases differ in the distances and angles between molecules or atoms and therefore the energy that binds them together. In solids, the structure is nearly rigid; in liquids, molecules or atoms move around each other but do not move apart; and in gases, molecules or atoms move almost independently of each other and are mostly far apart.

KEY
SE = Student Edition
ATE = Annotated Teacher Edition

Block 3 *(45 minutes)*
FOCUS *5 minutes*

❑ **Bellringer,** ATE (GENERAL). This activity has students list terms with the prefixes *inter-* and *intra-*.

MOTIVATE *10 minutes*

❑ **Discussion,** ATE (GENERAL). Lead a discussion to help students derive the meaning of the prefixes *inter-* and *intra-* using the lists they compiled in the Bellringer activity. Then help students define the term *intermolecular forces*.

TEACH *30 minutes*

❑ **Observation Lab: Constructing a Heating/Cooling Curve** Chapter Resource File (GENERAL). In this lab students collect heating/cooling data and from the data construct a heating/cooling curve.

❑ **Using the Table,** ATE (GENERAL). Point out that HF does not fit the pattern shown in Table 3.

❑ **Using the Figure,** ATE (GENERAL). Have students compare the strength of hydrogen bonds and ionic bonds in DNA.

HOMEWORK

❑ **Reading Skill Builder,** ATE (BASIC). Have students make a table in which they list the different types of intermolecular forces according to how each forms, their relative strengths, and examples of the types of molecules that exhibit the force.

OTHER RESOURCES

❑ **Skill Builder,** ATE (ADVANCED). With water's low molecular mass, it would exist on Earth as a gas if it did not contain hydrogen bonds. Have students write a story about what they think Earth would be like if hydrogen bonds did not exist in water.

❑ **go.hrw.com**

❑ **www.scilinks.org**

Block 4 *(45 minutes)*

TEACH *30 minutes*

❑ **Transparency,** Ice and Water. (GENERAL) This master shows how hydrogen bonding in water results in open hexagonal crystal structure in ice.

❑ **Group Activity,** ATE (GENERAL). This activity has students compare the density of ice and liquid water.

❑ **Transparency,** Temporary Dipoles. (GENERAL) This master shows how temporary dipoles cause forces of attraction between molecules.

❑ **CBL™ Probware Lab, Evaporation and Intermolecular Atrractions.** Chapter Resource File (ADVANCED). In this lab, students will monitor temperature changes of a variety of substances as each evaporates off of a temperature probe. They will then relate the change in temperature to each substance's molecular structure and presence of intermolecular forces.

CLOSE *10 minutes*

❑ **Reteaching,** ATE (BASIC). Students write an outline using the main headings "Dipole-dipole forces" and "London forces."

❑ **Quiz,** ATE (GENERAL). This assignment has students answer questions about the concepts in this lesson.

❑ **Assessment Worksheet: Section Quiz** (GENERAL)

HOMEWORK

❑ **Homework,** ATE (BASIC). Students construct a concept map.

❑ **Skills Worksheet: Concept Review** (GENERAL)

❑ **Section Review,** SE (GENERAL). Assign items 1–12.

OTHER RESOURCES

❑ **Skill Builder,** ATE (ADVANCED). Have students research the work of Johannes van der Waals.

❑ **go.hrw.com**

❑ **www.scilinks.org**

Lesson Plan

Section: Energy of State Changes

Pacing

Regular Schedule	with lab(s): NA	without lab(s): 2 days
Block Schedule	with lab(s): NA	without lab(s): 1 day

Objectives

1. Define the molar enthalpy of fusion and the molar enthalpy of vaporization of substances, and identify them for a particular substance by using a heating curve.

2. Describe how the enthalpy and entropy of a substance relate to the substance's state.

3. Predict whether a state change will take place using Gibbs energy.

4. Calculate melting point and boiling point by using enthalpy and entropy.

5. Explain how pressure affects the entropy of a gas and affects changes between the liquid and vapor states.

National Science Education Standards Covered
UNIFYING CONCEPTS AND PROCESSES

UCP 1 Systems, order, and organization

UCP 3 Change, constancy, and measurement

UCP 5 Form and function

> **KEY**
> **SE** = Student Edition
> **ATE** = Annotated Teacher Edition

Block 5 *(45 minutes)*
FOCUS *10 minutes*

❑ **Bellringer,** ATE (GENERAL). Students randomly disperse and arrange objects and relate the activites to entropy and states of matter.

MOTIVATE *10 minutes*

❑ **Demonstration,** ATE (GENERAL). This demonstration illustrates a change of state.

TEACH *25 minutes*

❑ **Using the Figure,** ATE (GENERAL). Use Figures 2 and 17 to discuss entropy, energy, and temperature.

❑ **Demonstration,** ATE (GENERAL). This demonstration illustrates a change of state.

❑ **Using the Table,** ATE (GENERAL). Have students compare the enthalpies and entropies of fusion with the enthalpies and entropies of vaporization.

HOMEWORK

❑ **Section Review,** SE (GENERAL). Assign items 1–6.

OTHER RESOURCES

❑ **Demonstration,** ATE (GENERAL). This demonstration illustrates how a piece of paper soaked in ethanol will not burn.

❑ **go.hrw.com**

❑ **www.scilinks.org**

Block 6 *(45 minutes)*

TEACH *35 minutes*

❑ **Demonstration,** ATE (GENERAL). This demonstration illustrates the synthesis of copper(II) sulfate.

❑ **Sample Problem A: Calculating Melting and Boiling Points,** SE (GENERAL). This problem demonstrates how to calculate melting and boiling points of a substance.

❑ **Group Activity,** ATE (GENERAL). Have small groups of students summarize each paragraph under the heading "Enthalpy and Entropy Determine State."

CLOSE *10 minutes*

❑ **Reteaching,** ATE (BASIC). Have students explain why a cup catches fire and burns but a cup filled with water does not.

❑ **Quiz,** ATE (GENERAL). This assignment has students answer questions about the concepts in this lesson.

❑ **Assessment Worksheet: Section Quiz** (GENERAL)

HOMEWORK

❑ **Practice Sample Problems A,** SE (GENERAL). Calculating Melting and Boiling Points of a Sample. Assign items 1–3.

❑ **Homework,** ATE (GENERAL). This assignment provides additional practice problems using calculating melting and boiling points of a sample, like those in Practice Problem A.

❑ **Skills Worksheet: Concept Review** (GENERAL)

❑ **Section Review,** SE (GENERAL). Assign items 7–13.

OTHER RESOURCES

❑ **go.hrw.com**

❑ **www.scilinks.org**

Lesson Plan

Section: Phase Equilibrium

Pacing

Regular Schedule   with lab(s): NA   without lab(s): 2 days
Block Schedule   with lab(s): NA   without lab(s): 1 day

Objectives

1. Identify systems that have multiple phases, and determine whether they are at equilibrium.

2. Understand the role of vapor pressure in changes of state between a liquid and a gas.

3. Interpret a phase diagram to identify melting points and boiling points.

National Science Education Standards Covered
UNIFYING CONCEPTS AND PROCESSES

UCP 1 Systems, order, and organization

UCP 3 Change, constancy, and measurement

UCP 5 Form and function

> **KEY**
> **SE** = Student Edition
> **ATE** = Annotated Teacher Edition

Block 7 *(45 minutes)*
FOCUS *5 minutes*

❑ **Bellringer,** ATE (GENERAL). Students oberve a wire pass through a large block of ice without cutting into it.

MOTIVATE *10 minutes*

❑ **Discussion,** ATE (GENERAL). Invite students to explain what they observed in the Bellgringer activity.

TEACH *30 minutes*

❏ **Demonstration,** ATE (GENERAL). This demonstration illustrates boiling water in a vacuum.

❏ **Transparency,** Energy Distribution of Gas Molecules at Different Temperatures. (GENERAL) This transparency master illustrates the energy distribution of gas molecules at different temperatures. (Figure 23)

❏ **Using the Figure,** ATE (GENERAL). Use this figure to discuss Figure 24. Have students answer a list of provided questions in reference to Figure 24.

❏ **Transparency,** Phase Diagram for H_2O. (GENERAL) This master shows the physical states of water at different temperatures and pressures.

HOMEWORK

❏ **Reading Skill Builder,** ATE (BASIC). Have students use self-sticking notes to annotate their textbook with questions as they read this section.

❏ **Section Review,** SE (GENERAL). Assign items 1–5.

OTHER RESOURCES

❏ **Misconception Alert,** ATE (GENERAL). Use this feature to compare the triple point to the apex of a pyramid.

❏ **Focus on Graphing,** SE (GENERAL).

❏ **go.hrw.com**

❏ **www.scilinks.org**

Block 8 *(45 minutes)*

TEACH *35 minutes*

❏ **Transparency,** Phase Diagram for CO_2. (GENERAL) This master shows the physical states of CO_2 at different temperatures and pressures.

❏ **Using the Figure,** ATE (GENERAL). Use this figure to have students confirm the values in Sample Problem B.

❏ **Group Activity,** ATE (BASIC). Students devise a game that involves interpreting a phase diagram.

❏ **Sample Problem B: How to Draw a Phase Diagram,** SE (GENERAL). This problem demonstrates how to draw a phase diagram.

CLOSE *10 minutes*

❏ **Reteaching,** ATE (BASIC). Students fill in the words and number missing on a blank phase diagram.

❏ **Quiz,** ATE (GENERAL). This assignment has students answer questions about the concepts in this lesson.

❏ **Assessment Worksheet: Section Quiz** (GENERAL)

HOMEWORK

❑ **Practice Sample Problems B,** SE (GENERAL). How to Draw a Phase Diagram. Assign items 1a–e.

❑ **Homework,** ATE (GENERAL). This assignment has students practice drawing phase diagrams. (Practice Problem B)

❑ **Skills Worksheet: Concept Review** (GENERAL)

❑ **Section Review,** SE (GENERAL). Assign items 6–12.

OTHER RESOURCES

❑ **Activity,** ATE (GENERAL). Students compare freeze dried foods with sun dried foods.

❑ **go.hrw.com**

❑ **www.scilinks.org**

END OF CHAPTER REVIEW AND ASSESSMENT RESOURCES

❑ **Mixed Review,** SE (GENERAL).

❑ **Alternate Assessment,** SE (GENERAL).

❑ **Technology and Learning,** SE (GENERAL).

❑ **Standardized Test Prep,** SE (GENERAL).

❑ **Assessment Worksheet: Chapter Test** (GENERAL)

❑ **Test Item Listing for ExamView® Test Generator**

Quick Lab

Wetting a Surface

OBJECTIVES

Test the wetting ability of water on several surfaces.

MATERIALS

- detergent, diluted with water
- glass plate
- plastic plate
- steel plate
- toothpick
- water

Always wear safety goggles, gloves, and a lab apron to protect your eyes and clothing. If you get a chemical in your eyes, immediately flush the chemical out at the eyewash station while calling to your teacher. Know the location of the emergency lab shower and eyewash station and the procedures for using them.

Procedure

1. Wash plastic, steel and glass plates well by using dilute detergent and rinse them completely. Do not touch the clean surfaces.

2. Using a toothpick, put a small drop of water on each plate. Observe the shape of the drops from the side.

Analysis

1. On which surface does the water spread the most?

   glass

2. On which surface does the water spread the least?

   plastic

3. What can you conclude about the adhesion of water for plastic, steel and glass?

   Water has a greater adhesion for glass than steel and for steel than plastic.

4. Explain your observations in terms of wetting.

   Water wets glass more than it wets steel or plastic.

Skills Practice Lab

Viscosity of Liquids

Teacher Notes

TIME REQUIRED One 45-minute lab period

SKILLS ACQUIRED
Collecting data
Communicating
Experimenting
Identifying patterns
Inferring
Interpreting
Organizing and analyzing data

RATING
Easy ← 1   2   3   4 → Hard

Teacher Prep–3
Student Set-Up–3
Concept Level–3
Clean Up–3

THE SCIENTIFIC METHOD

Make Observations Students collect calorimetry data using a variety of metals.

Analyze the Results Analysis questions 1 to 9

Draw Conclusions Analysis question 7 and Conclusions questions 10 to 16

Communicate the Results Analysis questions 1, 4, and 7 and Conclusions questions 11, 15, and 16

MATERIALS

For the oil samples, collect the grades of oil indicated in the following table, and label them with the letters *A–F*. It does not matter which type of oil gets labeled with which letter, provided you keep a record. The sample data and calculations shown in these teacher's notes are based on the following table.

Oil type	Letter
SAE-10	C
SAE-20	A
SAE-30	D
SAE-40	F
SAE-50	B
SAE-60	E

The oil samples should be reused from class to class to avoid the need to dispose of them.

Viscosity of Liquids *continued*

Once the pipet viscosimeters have been made and labeled, they can be reused from class to class. For best results, use only one grade of oil in each pipet viscosimeter.

It may help students complete the lab in a single lab period if you provide two sets of samples in test tubes. This way, students can be cooling and heating the different sets at the same time.

SAFETY CAUTIONS

Safety goggles, gloves, and a lab apron must be worn at all times to provide protection for the eyes and clothing.

Tie back long hair and loose clothing.

Read all safety cautions, and discuss them with your students.

Remind students to use a test-tube holder when removing test tubes from the warm-water bath.

The oils used in this lab should not be heated above 60°C.

Remind students of the following safety precautions:

- Always wear safety goggles, gloves, and a lab apron to protect your eyes and clothing. If you get a chemical in your eyes, immediately flush the chemical out at the eyewash station while calling to your teacher. Know the location of the emergency lab shower and the eyewash stations and procedures for using them.

- Do not touch any chemicals. If you get a chemical on your skin or clothing, wash the chemical off at the sink while calling to your teacher. Make sure you carefully read the labels, and follow the precautions on all containers of chemicals that you use. If there are no precautions stated on the label, ask your teacher what precautions you should follow. Do not taste any chemicals or items used in the laboratory. Never return leftovers to their original containers; take only small amounts to avoid wasting supplies.

- Pins are sharp; use them with care to avoid cutting yourself or others.

- When using a Bunsen burner, confine long hair and loose clothing. Do not heat glassware that is broken, chipped, or cracked. Use tongs or a test tube holder to handle heated glassware and other equipment. Because oil tested in this lab is flammable, it should never be heated directly over a flame. Instead use a hot-water bath, and never heat it above 60°C.

- Call your teacher in the event of a spill. Spills should be cleaned up promptly, according to your teacher's directions.

- Never put broken glass in a regular waste container. Broken glass should be disposed of properly.

Viscosity of Liquids *continued*

DISPOSAL

Set out twelve disposal containers, one for each of the six types of oil labeled *A* through *F* and for each of the six different pipets labeled *A* through *F*. Students in successive periods can reuse the labeled beakers, test tubes, and pipets. There is no need for a full-scale cleanup of all equipment until the end of the class day. The oil can be reused class after class and year after year. If the oil is cleaned up with paper towels, they can be disposed of only in a landfill designated for hazardous waste.

TECHNIQUES TO DEMONSTRATE

Be sure to demonstrate exactly how to prepare the pipet viscosimeter. Show students how to control the flow by covering the hole with a finger. It may take several trials before students have a consistent technique that allows measurements to be made. If students have trouble completing the lab in the time allotted, reduce the number of trials run at each temperature.

TIPS AND TRICKS

Make certain that the concepts of intermolecular bonding and hydrocarbon chain length discussed in the textbook are understood. Point out that the larger the molecule, the stronger the London forces can become.

Later, the technique used here of moving the finger away from a hole to allow liquid to flow out of the viscosimeter can be used as an object lesson to help students understand the role of atmospheric pressure. When a finger is held over the hole, the atmospheric pressure outside the pipet holds the fluid inside the stem.

Name _____ Class _____ Date _____

Skills Practice Lab

OBSERVATION

Viscosity of Liquids

SITUATION

You have been contacted by an automotive service shop that received a shipment of bulk containers of motor oil. The containers had been shipped by freight train, but several boxcars had leaky roofs. As a result, the labels peeled off the cans. Before the shop uses this oil in cars, the service technicians must match up the cans with the types of oil that were listed on the shipping invoice, based on the viscosity and the SAE rating of the oils.

BACKGROUND

Viscosity is the measurement of a liquid's resistance to flow. Several factors contribute to viscosity. Liquids with high intermolecular forces tend to be very viscous. For example, glycerol has a high viscosity because of its tendency to form many hydrogen bonds. For other molecules, such as oils, the longer the chain length of the molecule, the more viscous they are. The longer chains not only provide greater surface area for intermolecular attractions, but also can be intertwined more easily. For example, gasoline, which contains molecules that are chains of three to eight carbon atoms, is much less viscous than grease, which usually contains molecules with about 20 to 25 carbon atoms.

The Society of Automotive Engineers rates lubricating oils according to their comparative viscosities. These numerical values, called *SAE ratings*, range from SAE-10 (low viscosity) to SAE-60 (high viscosity) for oils typically used in combustion engines such as those in automobiles and trucks. The ratings are achieved with an instrument called a *viscosimeter*, which has a small capillary tube opening. The amount of time for a specific amount of motor oil to flow through the opening is a measure of viscosity. The less viscous oils flow through in a shorter time than the more viscous oils do.

PROBLEM

To match the correct oil sample to its SAE rating, you will need to do the following.

- Make your own viscosimeter from a pipet.
- Measure the relative viscosities of several oils by timing the oil as it flows through your viscosimeter.
- Measure mass and volume of each oil to calculate density.
- From the measurements, infer which labels belong on the containers of oil.

OBJECTIVES

Demonstrate proficiency in comparing the viscosity of various liquids under identical test conditions.

Construct a small viscosimeter.

Measure flow time of various single-weight oils.

Measure the mass and volume of the oils to calculate density.

Calculate the relative viscosity of the oils.

Graph experimental data.

Compare viscosities and densities to determine the SAE rating of each oil.

MATERIALS

- beakers, 400 mL (2)
- beakers, 50 mL (7)
- distilled water
- gloves
- graduated cylinder, 10 mL
- ice
- lab apron
- oil samples, 10 mL (6)
- pin, straight
- pipets, thin-stem (7)
- ruler, metric
- safety goggles
- stopwatch or clock with second hand
- test-tube holder
- test-tube rack
- test tubes, small (7)
- wax pencil

Bunsen burner option
- Bunsen burner and related equipment
- ring stand and ring
- wire gauze with ceramic center

Hot plate option
- hot plate

Thermometer option
- thermometer, nonmercury
- thermometer clamp

Probe option
- thermistor probe

Always wear safety goggles, gloves, and a lab apron to protect your eyes and clothing. If you get a chemical in your eyes, immediately flush the chemical out at the eyewash station while calling to your teacher. Know the location of the emergency lab shower and eyewash station and the procedures for using them.

Do not touch any chemicals. If you get a chemical on your skin or clothing, wash the chemical off at the sink while calling to your teacher. Make sure you carefully read the labels and follow the precautions on all containers of chemicals that you use. If there are no precautions stated on the label, ask your teacher what precautions to follow. Do not taste any chemicals or items used in the laboratory. Never return leftovers to their original container; take only small amounts to avoid wasting supplies.

Viscosity of Liquids *continued*

 Do not heat glassware that is broken, chipped, or cracked. Use tongs or a hot mitt to handle heated glassware and other equipment because hot glassware does not always look hot.

When using a Bunsen burner, confine long hair and loose clothing. If your clothing catches on fire, WALK to the emergency lab shower and use it to put out the fire. Because the oil tested in this lab is flammable, it should never be heated directly over a flame. Instead, use a hot-water bath, and never heat it above 60°C.

When heating a substance in a test tube, the mouth of the test tube should point away from where you and others are standing. Watch the test tube at all times to prevent the contents from boiling over.

Pins are sharp; use with care to avoid cutting yourself or others.

Procedure
PART 1–PREPARATION

1. Put on safety goggles, gloves, and a lab apron.

2. With a wax pencil, label each 50 mL beaker-test tube-pipet set with the name of one oil sample (*A, B, C, D, E,* or *F*). Label an additional set H_2O.

3. Place two marks 2.0 cm apart on the side of the bulb of the pipet, as shown in **Figure 1**. The top mark will be the starting point and the lower mark will be the endpoint.

4. Carefully make a small hole in the top of the bulb of each pipet with the pin, as shown in **Figure 1**. Be sure the hole is well above the marks you made on the side of the pipet bulb. Make the hole the same size for each pipet by putting in the pin the same way for each one. You will control the flow of oil with your finger and this hole.

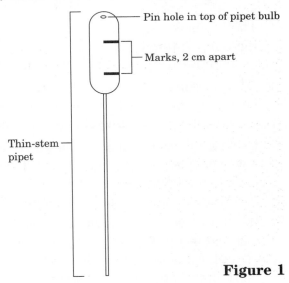
Figure 1

PART 2–TECHNIQUE

5. Measure the masses of all seven 50 mL beakers. Record them in your data table.

6. Pour about 5.0 mL of distilled water into the graduated cylinder. Measure and record the volume to the nearest 0.1 mL, and pour it into the H_2O beaker.

7. Measure and record the mass of the H_2O beaker with water in your data table.

8. Squeeze the H_2O pipet bulb and fill the pipet with distilled water to above the top line. After it is full, place your finger over the pin hole. Place the pipet over the H_2O beaker, lift your finger off the hole, and allow the liquid to flow into the beaker until the meniscus is even with the top line on the pipet bulb. Cover the hole promptly when the water reaches this point. Several practice trials may be necessary.

9. One member of the lab group should hold the pipet with a finger over the pinhole, and the other should use a clock with a second hand or a stopwatch to record precise time intervals. Hold the pipet over the H_2O beaker. When the timer is ready, remove your finger from the pinhole, and allow the liquid to flow into the beaker until it reaches the bottom line on the pipet bulb. Record the time elapsed to the nearest 0.1 s in your data table in the section for room temperature. (If you do not have a stopwatch, measure the time elapsed to the nearest 0.5 s.) It may take several practice trials to master the technique.

10. Repeat **steps 6–9** with each oil, using the appropriately labeled pipets and beakers. You should perform several trials for each oil and for water to obtain consistent results. Clean the graduated cylinder after the last trial for each oil.

11. Using one of the 400 mL beakers, make an ice bath. Fill the test tubes to within 1.0 cm of the top with the appropriate oil or distilled water. Cool the samples for 5–8 min so that they are at a temperature between 0°C and 10°C. The key is that all of the samples must be at the same temperature. Measure the temperature of the water sample to the nearest 0.1°C with a thermometer or a thermistor probe and record it below your data table.

12. Repeat **steps 8–9** with each of the cooled samples. Be sure to use the pipets and 50 mL beakers designated for each oil or distilled water. Record the volume, mass, and time elapsed for each trial in your data table.

13. Using a Bunsen burner or a hot plate and the second 400 mL beaker, prepare a warm-water bath with a temperature between 35°C and 45°C. If you measure the temperature with a thermometer, use a thermometer clamp attached to a ring stand to hold the thermometer in the water.

14. Refill the test tubes to within 1.0 cm of the top with the appropriate oil or distilled water. Place these test tubes into the warm-water bath, and allow the oil and water to warm. Measure the temperature of the water sample with a thermometer or a thermistor probe when you remove the samples and record it below **Table 1.**

15. Repeat **steps 8–9** with the warm samples. Record the volume, mass, and time elapsed for each trial in your data table.

16. Your instructor will have set out twelve disposal containers; six for the six types of oil and six for the pipets. **Do not pour oil down the sink. Do not put the oil or oily pipets in the trash can.** The distilled water may be poured down the sink. The test tubes should be washed with a mild detergent and rinsed. Always wash your hands thoroughly after cleaning up the area and equipment.

Viscosity of Liquids continued

TABLE 1 FLOW TIMES OF THE OILS

Sample	Beaker mass (g)	Total mass (g)	Volume (mL)	Trial 1— cool (s)	Trial 2— cool (s)	Trial 3— cool (s)
A	47.06	51.51	5.0	31.1	31.2	31.3
B	46.81	51.45	5.0	148.5	148.4	148.6
C	47.27	51.63	5.0	20.3	20.2	20.1
D	48.04	52.55	5.0	43.5	43.7	43.6
E	47.53	52.22	5.0	190.5	190.4	190.3
F	47.60	52.15	5.0	74.5	74.6	74.4
H_2O	46.92	51.92	5.0	1.6	1.6	1.6

Sample	Trial 1— room temp. (s)	Trial 2— room temp. (s)	Trial 3— room temp. (s)	Trial 1— warm (s)	Trial 2— warm (s)	Trial 3— warm (s)
A	14.7	14.6	14.5	7.3	7.2	7.1
B	51.3	51.4	51.2	18.3	18.4	18.2
C	10.5	10.3	10.4	5.5	5.3	5.4
D	20.9	21.1	21.0	9.8	9.9	9.7
E	73.6	73.7	73.5	25.5	25.3	25.4
F	31.7	31.6	31.8	11.9	11.7	11.8
H_2O	1.6	1.5	1.6	1.5	1.6	1.5

cool temperature: 4°C _____

room temperature: 20°C _____

warm temperature: 40°C _____

Analysis

1. Organizing Data Determine the density of each sample.

A: $\dfrac{4.45 \text{ g}}{5.0 \text{ mL}} = 0.89 \text{ g/mL}$

E: $\dfrac{4.69 \text{ g}}{5.0 \text{ mL}} = 0.94 \text{ g/mL}$

B: $\dfrac{4.64 \text{ g}}{5.0 \text{ mL}} = 0.93 \text{ g/mL}$

F: $\dfrac{4.55 \text{ g}}{5.0 \text{ mL}} = 0.91 \text{ g/mL}$

C: $\dfrac{4.36 \text{ g}}{5.0 \text{ mL}} = 0.87 \text{ g/mL}$

H_2O: $\dfrac{5.00 \text{ g}}{5.0 \text{ mL}} = 1.0 \text{ g/mL}$

D: $\dfrac{4.51 \text{ g}}{5.0 \text{ mL}} = 0.90 \text{ g/mL}$

Name _____ Class _____ Date _____

2. Organizing Data Find the average flow time for each sample at each temperature.

Sample	Avg. time (s) cool	Avg. time (s) room temp.	Avg. time (s) warm
A	31.2	14.6	7.2
B	148.5	51.3	18.3
C	20.2	10.4	5.4
D	43.6	21.0	9.8
E	190.4	73.6	25.4
F	74.5	31.7	11.8
H_2O	1.6	1.6	1.5

3. Analyzing Information Calculate the relative viscosity of your samples at room temperature by applying the following formula. The values for the absolute viscosity of water are in units of centipoises (cp). A centipoise is equal to 0.01 g/cm·s.

$$\text{relative viscosity}_{oil} = \frac{\text{density}_{oil} \times \text{time elapsed}_{oil} \times \text{viscosity}_{H_2O}}{\text{density}_{H_2O} \times \text{time elapsed}_{H_2O}}$$

Temperature (°C)	Absolute Viscosity for H_2O (cp)
18	1.053
20	1.002
22	0.955
24	0.911
25	0.890
26	0.870
28	0.833

relative viscosity$_A$: $\dfrac{0.89 \text{ g/mL} \times 14.6 \text{ s} \times 1.002 \text{ cp}}{1.0 \text{ g/mL} \times 1.6 \text{ s}}$ = 8.1 cp

relative viscosity$_B$: $\dfrac{0.93 \text{ g/mL} \times 51.3 \text{ s} \times 1.002 \text{ cp}}{1.0 \text{ g/mL} \times 1.6 \text{ s}}$ = 30. cp

relative viscosity$_C$: $\dfrac{0.87 \text{ g/mL} \times 10.4 \text{ s} \times 1.002 \text{ cp}}{1.0 \text{ g/mL} \times 1.6 \text{ s}}$ = 5.7 cp

relative viscosity$_D$: $\dfrac{0.90 \text{ g/mL} \times 21.0 \text{ s} \times 1.002 \text{ cp}}{1.0 \text{ g/mL} \times 1.6 \text{ s}}$ = 12 cp

relative viscosity$_E$: $\dfrac{0.94 \text{ g/mL} \times 73.6 \text{ s} \times 1.002 \text{ cp}}{1.0 \text{ g/mL} \times 1.6 \text{ s}}$ = 43 cp

relative viscosity$_F$: $\dfrac{0.91 \text{ g/mL} \times 31.7 \text{ s} \times 1.002 \text{ cp}}{1.0 \text{ g/mL} \times 1.6 \text{ s}}$ = 18 cp

Conclusions

4. Inferring Conclusions According to the invoice, the service station was supposed to receive equal amounts of SAE-10, SAE-20, SAE-30, SAE-40, SAE-50, and SAE-60 oil. Given that the oils with the lower SAE ratings have lower relative viscosities, infer which oil samples correspond to the SAE ratings indicated.

<u>If the oils were labeled as in the table in the Materials section, students</u>

<u>should identify the following matches. Otherwise, refer to your notes about</u>

<u>which oil was labeled with which letter.</u>

Letter	Rel. viscosity (cp)	Oil type
A	8.1	SAE-20
B	30	SAE-50
C	5.7	SAE-10
D	12	SAE-30
E	43	SAE-60
F	18	SAE-40

5. Organizing Information Prepare a graph with flow time at room temperature on the *y*-axis and SAE rating on the *x*-axis.

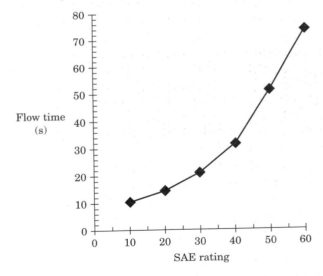

Viscosity of Liquids *continued*

6. Organizing Information Prepare a graph with density on the *y*-axis and SAE rating on the *x*-axis.

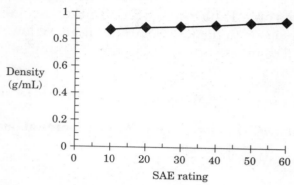

7. Organizing Information Prepare a graph with viscosity at room temperature on the *y*-axis and SAE rating on the *x*-axis.

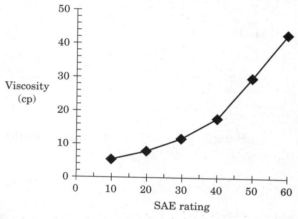

8. Organizing Information Prepare a graph with viscosity at room temperature on the *y*-axis and density on the *x*-axis.

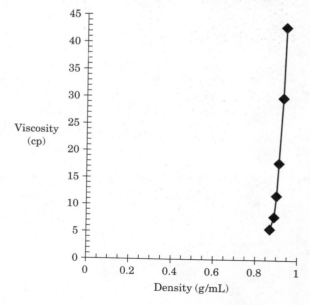

Viscosity of Liquids *continued*

9. Inferring Conclusions How does temperature affect the viscosity of each sample?

All of the oil samples are more viscous at low temperatures and less viscous
at high temperatures.

10. Interpreting Graphics Is there a relationship between density and viscosity?

The densities of the various oils do not vary by much. However, some
students may notice that as the density increases, so does the viscosity.

11. Interpreting Graphics What is the relationship between SAE rating and viscosity?

The lower the SAE rating, the lower the viscosity of the oil.

12. Interpreting Graphics What is the relationship between viscosity and flow time?

More viscous fluids take more time to flow.

Extensions

1. Predicting Outcomes Estimate what flow times you would measure at each temperature if you repeated the tests in this lab with SAE-35 oil.

Given the patterns detected so far, SAE-35 should have flow times between
those of SAE-30 and SAE-40. Answers should be near 59.0 s for 4°C, 26.4 s
for 20°C, and 10.8 s for 40°C.

Viscosity of Liquids *continued*

2. Relating Ideas Malcolm is trying to get the last of the pancake syrup out of a bottle. What can he do to make the syrup come out of the bottle faster? Explain how your plan will take advantage of viscosity.

<u>If Malcolm heats the bottle of pancake syrup, the syrup will be less viscous</u>

<u>and should flow more quickly.</u>

_____

_____

3. Research and Communication Contact a manufacturer of lubrication products such as Valvoline or Pennzoil, and write a short paper on the development and properties of the oils used in this investigation.

<u>Student answers will vary. Be certain students realize that the different</u>

<u>properties of the different grades of oil are a result of the different</u>

<u>combinations of ingredients used for each grade.</u>

_____

Constructing a Heating/Cooling Curve

Teacher Notes

TIME REQUIRED One 45-minute lab period

SKILLS ACQUIRED
Collecting data
Communicating
Experimenting
Identifying patterns
Inferring
Interpreting
Organizing and analyzing data

RATING

Easy ◄— 1 2 3 4 —► Hard

Teacher Prep–3
Student Set-Up–4
Concept Level–3
Clean Up–3

THE SCIENTIFIC METHOD

Make Observations Students collect temperature data as a substance is cooled and heated, then they create heating and cooling curves.

Analyze the Results Analysis questions 1 to 5 and Conclusions questions 6 and 7

Draw Conclusions Conclusions questions 8 to 10 and Analysis question 4 and 5

Communicate the Results Analysis questions 2 to 4 and Conclusions questions 6 to 10

MATERIALS

This lab will go more quickly if several wire stirrers are prepared in advance. Cut 25 cm lengths of wire. (Any gauge will do provided it is easily bent.) Make a loop at one end that has a 1 cm diameter. Bend the wire where it attaches to the loop so that the loop is perpendicular to the rest of the wire. Bend the top point of the wire over in a small loop to use as a handle.

Fill test tubes with about 15.0 g of $Na_2S_2O_3 \cdot 5H_2O$. The solid can be reused several times. Measure out one 15 g sample, completely transfer it to a test tube, and fill the remaining test tubes to the same level.

Be sure to set out a wide-mouthed bottle containing several small crystals of $Na_2S_2O_3 \cdot 5H_2O$ for students to use as seed crystals.

Other test-tube clamps may be used in place of the three-fingered one.

SAFETY CAUTIONS

Safety goggles, gloves, and a lab apron must be worn at all times.

Read all safety cautions, and discuss them with your students.

Make sure the iron rings are large enough to hold a 600 mL beaker.

Constructing a Heating/Cooling Curve *continued*

Remind students of the following safety precautions:

• Always wear safety goggles, gloves, and a lab apron to protect your eyes and clothing. If you get a chemical in your eyes, immediately flush the chemical out at the eyewash station while calling to your teacher. Know the location of the emergency lab shower and the eyewash stations and procedures for using them.

• Do not touch any chemicals. If you get a chemical on your skin or clothing, wash the chemical off at the sink while calling to your teacher. Make sure you carefully read the labels, and follow the precautions on all containers of chemicals that you use. If there are no precautions stated on the label, ask your teacher what precautions you should follow. Do not taste any chemicals or items used in the laboratory. Never return leftovers to their original containers; take only small amounts to avoid wasting supplies.

• Call your teacher in the event of a spill. Spills should be cleaned up promptly, according to your teacher's directions.

• Never put broken glass in a regular waste container. Broken glass should be disposed of properly.

DISPOSAL

Remelt the sodium thiosulfate pentahydrate in a water bath, pour all of the liquid in a wide-mouth reagent jar, cool to room temperature, cover, and label for reuse. It will be necessary to pulverize the crystals into smaller chunks before reusing them.

TECHNIQUES TO DEMONSTRATE

Students find it particularly awkward to manipulate the setups, so a quick run-through of the steps may prove beneficial: assembling and operating the hot-water bath, positioning the thermometer in the solid, raising and lowering and exchanging the beakers, and when to begin and end timing.

TIPS AND TRICKS

Remind students that thermometer bulbs must not rest on the bottom of test tubes or beakers and that the thermometers must never be used to stir anything. The most accurate temperature readings are made when the thermometer is vertical and the line of sight is horizontal, not angled.

Point out the purpose of the initial quick melt. Emphasize that only one temperature is taken during the quick melt. This provides a rough idea of where the melting and freezing point will occur. In the succeeding tests, more attention should be paid to the temperature readings near this quick-melt temperature, and repetitive readings should be expected.

Explain that the temperature of a pure substance remains constant during a phase change. Review what happens at the particle level during melting and freezing.

NOTE: The $Na_2S_2O_3 \cdot 5H_2O$ will undercool unless a seed crystal is added a few degrees above its freezing temperature. The temperature may still dip one or two degrees below the freezing temperature, but it will rise when the liquid is stirred.

Name _____ Class _____ Date _____

Skills Practice Lab

Constructing a Heating/Cooling Curve

SITUATION

Sodium thiosulfate pentahydrate, $Na_2S_2O_3 \cdot 5H_2O$, is produced by a local manufacturing firm and sold nationwide to photography shops, paper processing plants, and textile manufacturers. Purity is one condition of customer satisfaction, so samples of $Na_2S_2O_3 \cdot 5H_2O$ are taken periodically from the production line and tested for purity by an outside testing facility. Your company has been tentatively chosen because your proposal was the only one based on melting and freezing points rather than the more expensive titrations with iodine. To make the contract final, you must convince the manufacturing firm that you can establish accurate standards for comparison.

BACKGROUND

As energy flows from a liquid, its temperature drops. The entropy, or random ordering of its particles, also decreases until a specific ordering of the particles results in a phase change to a solid. If energy is being released or absorbed by a substance remaining at the same temperature, this is evidence that a dramatic change in entropy, such as a phase change, is occurring. Because all of the particles of a pure substance are identical, they all freeze at the same temperature, and the temperature will not change until the phase change is complete. If a substance is impure, the impurities will not lose energy in the same way that the rest of the particles do. Therefore, the freezing point will be somewhat lower, and there will be a range of temperatures instead of a single temperature.

PROBLEM

To evaluate the samples, you will need a heating/cooling curve for pure $Na_2S_2O_3 \cdot 5H_2O$ that you can use as a standard. To create and use this curve, you must do the following.

- Obtain a measured amount of pure $Na_2S_2O_3 \cdot 5H_2O$.

- Melt and freeze the sample, periodically recording the time and temperature.

- Graph the data to determine the melting and freezing points of pure $Na_2S_2O_3 \cdot 5H_2O$.

- Interpret the changes in energy and entropy involved in these phase changes.

- Verify the observed melting point against the accepted melting point found in reference data from two different sources.

- Use the graph to qualitatively determine whether there are impurities in a sample of $Na_2S_2O_3 \cdot 5H_2O$.

Constructing a Heating/Cooling Curve *continued*

OBJECTIVES

Observe the temperature and phase changes of a pure substance.

Measure the time needed for the melting and freezing of a specified amount of substance.

Graph experimental data and determine the melting and freezing points of a pure substance.

Analyze the graph for the relationship between melting point and freezing point.

Identify the relationship between temperature and phase change for a substance.

Infer the relationship between energy and phase changes.

Recognize the effect of an impurity on the melting point of a substance.

Analyze the relationship between energy, entropy, and temperature.

MATERIALS

- balance, centigram
- beaker tongs
- beakers, 600 mL (3)
- chemical reference books
- forceps
- gloves
- graph paper
- hot mitt
- ice
- lab apron
- $Na_2S_2O_3 \cdot 5H_2O$
- plastic washtub
- ring clamps (3)
- ring stands (2)
- ruler
- safety goggles

- stopwatch or clock with a second hand
- test-tube clamp
- test tube, Pyrex, medium
- thermometer clamp
- wire gauze with ceramic center (2)
- wire stirrer

Bunsen burner option

- Bunsen burner
- gas tubing
- striker

Hot plate option

- hot plate

Probe option

- thermistor probes (2)

Thermometer option

- thermometers, nonmercury (2)

Always wear safety goggles, gloves, and a lab apron to protect your eyes and clothing. If you get a chemical in your eyes, immediately flush the chemical out at the eyewash station while calling to your teacher. Know the location of the emergency lab shower and eyewash station and the procedures for using them.

Name _____ Class _____ Date _____

Constructing a Heating/Cooling Curve *continued*

Do not touch any chemicals. If you get a chemical on your skin or clothing, wash the chemical off at the sink while calling to your teacher. Make sure you carefully read the labels and follow the precautions on all containers of chemicals that you use. If there are no precautions stated on the label, ask your teacher what precautions to follow. Do not taste any chemicals or items used in the laboratory. Never return leftovers to their original container; take only small amounts to avoid wasting supplies.

Do not heat glassware that is broken, chipped, or cracked. Use tongs or a hot mitt to handle heated glassware and other equipment because hot glassware does not always look hot.

When using a Bunsen burner, confine long hair and loose clothing. If your clothing catches on fire, WALK to the emergency lab shower and use it to put out the fire.

When heating a substance in a test tube, the mouth of the test tube should point away from where you and others are standing. Watch the test tube at all times to prevent the contents from boiling over.

Procedure
PART 1–PREPARATION

1. Put on safety goggles, gloves, and a lab apron.

2. Fill two 600 mL beakers three-fourths full of tap water.

3. Heat water for a hot-water bath. If you are using a Bunsen burner, attach to a ring stand a ring clamp large enough to hold a 600 mL beaker. Adjust the height of the ring until it is 10 cm above the burner. Cover the ring with wire gauze. Set one 600 mL beaker of water on the gauze. If you are using a hot plate, rest the beaker of water directly on the hot plate.

4. Monitor the temperature of the water with a thermometer or a thermistor probe. Complete **steps 5–8** while the water is heating.

5. Cool the water for a cold-water bath. Fill a small plastic washtub with ice. Form a hole in the ice that is large enough for the second 600 mL beaker. Insert the beaker and pack the ice around it up to the level of the water in the beaker.

6. Bend the piece of wire into the shape of a stirrer, as shown in **Figure 1.** One loop should be narrow enough to fit into the test tube, yet wide enough to easily fit around the thermometer without touching it.

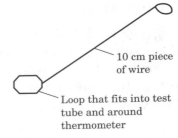

10 cm piece of wire

Loop that fits into test tube and around thermometer

Figure 1

Constructing a Heating/Cooling Curve *continued*

7. Prepare the sample. Assemble the test tube, thermometer, and stirrer, as shown in **Figure 2**. Attach the entire assembly to a second ring stand. Then, add enough $Na_2S_2O_3\cdot5H_2O$ crystals so that the test tube is about one-quarter full and the thermometer bulb is well under the surface of the crystals as shown in **Figure 3**.

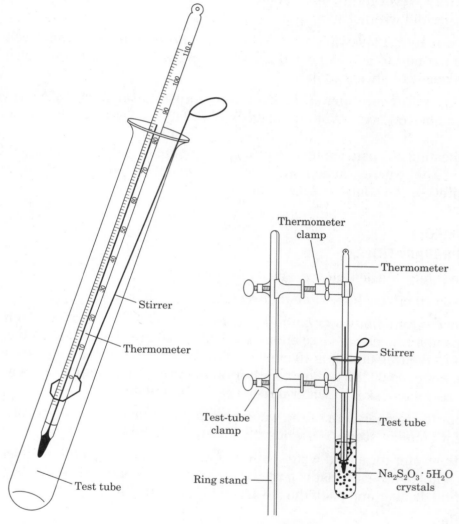

Figure 2   **Figure 3**

8. Set up the container for the hot-water bath as shown in **Figure 4**. Attach two ring clamps, one above the other, to the second ring stand beneath the test-tube assembly. Place a wire gauze with ceramic center on the lower ring. Set a third 600 mL beaker, which should be empty, on the gauze and raise the beaker toward the test-tube assembly until it surrounds nearly one-half of the tube's length. The beaker will pass through the ring clamp without gauze, and the test tube should not touch the bottom or sides of the beaker, as shown in **Figure 4** on the next page. The top clamp keeps the beaker from tipping when the beaker is filled with the hot water.

Constructing a Heating/Cooling Curve *continued*

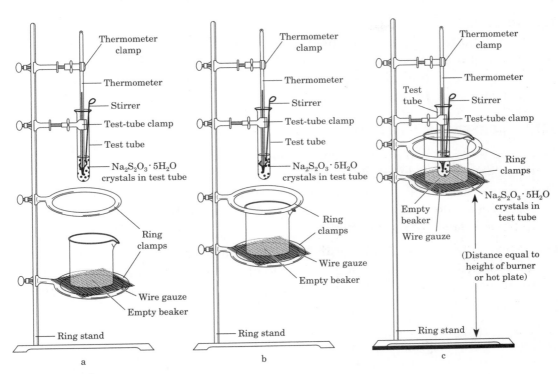

Figure 4

PART 2–MELTING A SOLID: QUICK TEST

9. Check the temperature of the water for the hot-water bath. When it is 85°C, turn off the burner or hot plate. If the temperature is already greater than 85°C, shut off the burner or hot plate, and add a few pieces of ice to bring the temperature down to 85°C. Then, using beaker tongs, remove the beaker of hot water from the burner. Using tongs or a hot mitt carefully pour the water into the empty beaker until the water level is well above the level of the solid inside the test tube. Set the empty beaker on the counter. You will use it again in step **20.**

10. Begin timing. The second the water is poured, one member of the lab group should begin timing, while the other reads the initial temperatures of the bath with one thermometer and sample with the other thermometer.

11. Occasionally stir the melting solid by gently moving the stirrer up and down. Be careful not to break the thermometer bulb. Monitor the temperature of the $Na_2S_2O_3 \cdot 5H_2O$ and the hot-water bath with separate thermometers or probes.

12. When the temperature of the liquid $Na_2S_2O_3 \cdot 5H_2O$ is approximately the same as that of the hot-water bath, stop timing. Note the final temperature of the liquid $Na_2S_2O_3 \cdot 5H_2O$ and the elapsed time. This temperature is the approximate melting point of your sample. Knowing this value can help you make the careful observations necessary to determine a more precise value.

Name _____ Class _____ Date _____

Constructing a Heating/Cooling Curve *continued*

13. Using a hot mitt, hold the beaker of hot water with one hand while using the other hand to gently loosen only the lower ring clamp enough so that the beaker of hot water can be lowered and removed. Remove the beaker of hot water, set it on the gauze above the burner, and let it reheat to 65°C while you perform **steps 14–20.**

PART 3—FREEZING A LIQUID

14. Set up the cold-water bath. Remove the beaker of cold water from the ice and place it on the ring with the gauze, well below the test tube. Steady the beaker with one hand while raising it until the level of the cold water is well above the level of the liquid inside the test tube. The test tube should not touch the bottom or sides of the beaker.

15. Begin timing. The second that the cold water is in place, one member of the lab group should begin timing, while the other reads the initial temperatures of the sample and the bath. Record the initial time and temperatures in the left half of **Table 1.** The starting temperature of the liquid should be near 80°C.

16. Monitor the cooling process. Measure and record the time and the temperature of the $Na_2S_2O_3 \cdot 5H_2O$ every 15 s in the left half of **Table 1.** Also record observations about the substance's appearance and other properties in the *Observations of cooling* column in **Table 1.** When the temperature reaches 50°C, use forceps to add one or two seed crystals of $Na_2S_2O_3 \cdot 5H_2O$ to the test tube.

17. Continue taking temperature readings every 15 s, stirring continuously, until a constant temperature is attained. (A temperature is constant if it is recorded at four consecutive 15 s intervals.) **Do not try to move the thermometer, thermistor probe, or stirrer when solidification occurs.**

18. Finish timing. Continue taking readings until the temperature of the solid differs from the temperature of the cold-water bath by 5°C.

19. Remove the cold-water bath. Grasp the beaker with one hand, carefully loosen its supporting ring clamp with the other hand, and lower the beaker of cold water away from the test tube. Remove the beaker from the ring and set it on the counter.

PART 3—MELTING A SOLID

20. Set up the container for the hot-water bath. Place the empty beaker from **step 9** on the ring and wire gauze. Steady the beaker as you raise it to surround the test tube as you did in **step 8,** but this time allow room for the Bunsen burner to be placed under the beaker.

Constructing a Heating/Cooling Curve *continued*

21. Fill the hot-water bath. Use the second thermometer or thermistor probe to check the temperature of the water for the hot-water bath. When it is 65°C, turn off the burner or hot plate. If the temperature is greater than 65°C, add a few pieces of ice to lower the temperature. Using tongs or a hot mitt, carefully pour the hot water into the empty beaker until the water level is well above the level of the solid inside the test tube. Set the empty beaker on the counter.

22. Begin timing. The second that the water is poured, one member of the lab group should begin timing while another reads initial temperatures of the water bath and the solid $Na_2S_2O_3 \cdot 5H_2O$. Record the solid's temperature in the right half of the data table. The starting temperature of the solid should be below 35°C.

23. Maintain the bath's temperature. Move the burner or hot plate under the hot-water bath and continue heating the water in the bath. Adjust the position and size of the flame or the setting of the hot plate so that the temperature of the hot-water bath remains between 60°C and 65°C.

24. Monitor the warming process. Record the temperature of the sample every 15 s. Use the stirrer, when it becomes free of the solid, to gently stir the contents of the test tube. Also record observations about the substance's appearance and other properties in the *Observations of warming* column in **Table 1.**

25. Continue taking readings until the temperature of the $Na_2S_2O_3 \cdot 5H_2O$ differs from that of the hot-water bath by 5°C.

26. Record the final temperature and the time.

27. Turn off the burner or hot plate.

28. Remove the thermometer or thermistor probe from the liquid $Na_2S_2O_3 \cdot 5H_2O$ and rinse it. Pour the $Na_2S_2O_3 \cdot 5H_2O$ from the test tube into the disposal container designated by your teacher. If you used a Bunsen burner, check to see that the gas valve is completely turned off. Remember to wash your hands thoroughly after cleaning up the lab area and all equipment.

Name _____ Class _____ Date _____

Constructing a Heating/Cooling Curve *continued*

TABLE 1 TIME AND TEMPERATURE DATA

Cooling Data			Warming Data		
Time (s)	Temp. (°C)	Observations of cooling	Time (s)	Temp. (°C)	Observation of warming
0	69.0		0	29.0	
15	61.0		15	31.5	
30	51.0	Seed crystal added	30	36.0	
45	47.0	Crystallization	45	39.5	
60	46.0		60	42.0	
75	48.2		75	43.5	
90	48.2		90	44.5	
105	48.2		105	45.5	
120	48.2		120	46.0	Melting starts
135	48.2		135	46.5	
150	48.2		150	47.0	
165	48.2		165	47.0	
180	48.2		180	47.2	
195	48.2		195	47.4	
210	48.2		210	47.5	
225	48.2		225	47.5	
240	48.0	Total solid	240	47.5	
255	47.5		255	48.0	
270	47.0		270	55.0	All solid melted
285	45.7		285	55.0	
300	44.7		300	58.0	
315	43.5		315	60.0	
330	41.5				
345	39.0				
360	36.2				
375	34.0				
390	32.0				

Name _____ Class _____ Date _____

Constructing a Heating/Cooling Curve *continued*

Analysis

1. **Organizing Data** Plot both the heating and cooling data on the same graph. Place time on the x-axis and temperature on the y-axis.

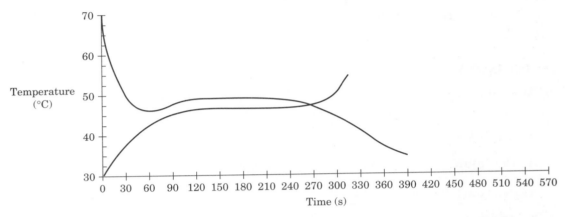

2. **Interpreting Graphics** Describe and compare the shape of the cooling curve with the shape of the heating curve.

   The cooling curve begins at an initial high temperature, slopes rapidly

   downward to a plateau at 48.2°C, and then continues to slope downward. The

   heating curve begins at an initially low temperature, slopes rapidly upward

   to a plateau, and then continues to slope upward. The two curves overlap in

   the plateau area.

3. **Interpreting Graphics** Locate the freezing and melting temperatures on your graph. Compare them and comment on why they have different names.

   The melting point is 48.2°C, while the freezing point is 47.5°C. They are

   close but should be the same. The difference in names reflects the phase

   change that is taking place and tells from which direction the phase change

   is being approached.

4. **Evaluating Methods** One purpose of the quick test for melting point is summarized in **step 12**. State this purpose and explain how it prepares you for **steps 17 and 24**.

   The quick melt provides a rough measure of the melting temperature. It tells

   the experimenter to pay close attention to time and temperature measure-

   ments near this point because the phase change plateau should occur.

5. Evaluating Data Compare your melting point with that found in references. What is your percent error?

from Merck Index: m.p. = 48°C when rapidly heated

$$\frac{48.2 - 48}{48} \times 100 = 1.67\%$$

Conclusions

6. Analyzing Information As the liquid cools, what is happening to the kinetic energy and the entropy of the following?

a. $Na_2S_2O_3 \cdot 5H_2O$

As the liquid cools, the kinetic energy and entropy decrease for the

$Na_2S_2O_3 \cdot 5H_2O$ particles.

b. the water bath

As the liquid cools, the kinetic energy and the entropy of the particles in the

water bath increase.

7. Analyzing Information What happened to the temperature of the sample from the time that freezing began until freezing was complete? Did the entropy of the sample increase, decrease, or stay the same?

The temperature remained relatively constant, but the entropy decreased

while the liquid was freezing into a solid.

8. Predicting Outcomes How would the quantity of the sample affect the time needed for the melting point test?

The larger the sample size, the longer the sample remains at the plateau

temperature.

9. Predicting Outcomes Would the quantity of the sample used to determine the melting point affect its outcome? (Hint: is melting point an extensive or intensive property?)

No, melting point is an intensive property, so it depends on the nature of the

substance being melted, not on the quantity.

Constructing a Heating/Cooling Curve *continued*

10. Interpreting Graphics

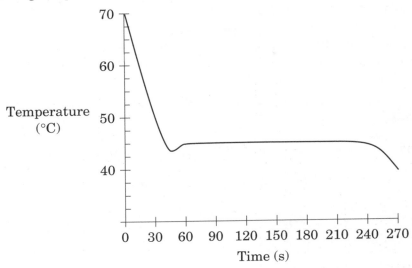

Examine the graph above, and compare it to your cooling curve. Would this sample of sodium thiosulfate pentahydrate be considered pure or impure? Sketch a line on the graph that represents your cooling data. If the curves are not identical, estimate the difference in melting points.

The substance is impure because the melting point is different; ΔT is about

3°C.

_____

Extensions

1. **Interpreting Graphics** Refer to the heating and cooling curves you plotted. For each portion of the curve, describe what happens to the energy and entropy of the substance.

   The energy and entropy decrease during cooling but increase during heating.

   The downward slope seems to indicate a large transfer of energy to the

   surroundings, so molecules are moving less rapidly; the upward slope indi-

   cates a large transfer from the surroundings, so molecular motion increases.

   The phase changes during plateaus represent times at which there are no

   changes in temperature and only small changes in molecular motion, even

   though substantial changes occur in energy and entropy.

Name _____ Class _____ Date _____

Constructing a Heating/Cooling Curve *continued*

2. **Applying Ideas** In northern climates, freezing rain is a driving hazard. When this occurs, warm air from a defroster is blown against the windshield of an automobile in order to restore visibility. It would be convenient to have a system that automatically turned the defroster blower system on and off as needed. A thermostat embedded in the windshield to detect outside temperature could be used to perform this function.

a. At what temperature should the thermostat be set to turn on the hot-air blower?

The thermostat should turn on the freezing-rain defroster when the

temperature is only a few degrees above freezing (0°C).

b. At what temperature should the thermostat be set to turn off the hot-air blower?

The defroster should be turned off when the temperature is warm enough

that the freezing rain should melt immediately, at approximately 20°C.

3. **Analyzing Methods** Will crystallization take place if no seed crystal is added? Why or why not?

Crystallization should eventually take place when the temperature is below

the freezing point. However, this process will not be as rapid as when the

cooling water bath and seed crystal are used.

4. **Designing Experiments** Explain the purpose of a water bath. Why wasn't distilled water necessary?

A water bath warms the test tube evenly. Distilled water is not necessary

because the water does not mix with the chemicals.

5. Organizing Ideas Which of the following word equations best represents the changes of phase taking place in the situations described in items **1** to **9** below? Place your answer in the space to the left of the numbered items.

a. solid + energy $\longrightarrow$ liquid

b. liquid $\longrightarrow$ solid + energy

c. solid + energy $\longrightarrow$ vapor

d. vapor $\longrightarrow$ solid + energy

___a___ **1.** ice melting at 0°C

___b___ **2.** water freezing at 0°C

__a, b__ **3.** a mixture of ice and water whose relative amounts remain unchanged

___c___ **4.** a particle escaping from a solid and becoming a vapor particle

___c___ **5.** solids, like camphor and naphthalene, subliming

___a___ **6.** snow melting

___b___ **7.** snow forming

___c___ **8.** dry ice subliming

___d___ **9.** dry ice forming

Evaporation and Intermolecular Attractions

Time Required

One lab period

Skills Acquired

- Collecting data
- Experimenting
- Organizing and analyzing data
- Interpreting
- Drawing real-world conclusions

The Scientific Method

- **Form a Hypothesis** In Procedure Steps 13 and 15 students predict the change in temperature based on molecular weight and collected data.

- **Analyze the results** In Analysis questions 1–5, students will analyze the data and results from their experiment.

- **Draw conclusions** In Conclusions questions 1–4, students will interpret the data and apply them to the objectives of the experiment.

Teacher's Notes

MATERIALS AND EQUIPMENT

- We recommend wrapping the probes with paper as described in the procedure. Wrapped probes provide more-uniform liquid amounts, and generally greater ΔT values, than bare probes. Chromatography paper, filter paper, and various other paper types work well.

- Snug-fitting rubber bands can be made by cutting short sections from a small rubber hose. Surgical tubing works well. Orthodontist's rubber bands are also a good size.

- Sets of the liquids can be supplied in 13 mm × 100 mm test tubes stationed in stable test-tube racks. This method uses very small amounts of the liquids. Alternatively, the liquids can be supplied in sets of small bottles kept for future use. Adjust the level of the liquids in the containers so it will be above the top edge of the filter paper.

- The temperature calibrations that are stored in the DataMate data-collection program will work fine for this experiment. No calibration is necessary for the temperature probes.

- The Vernier stainless steel temperature probe and CBL Temperature probe will plug directly into CH1 on the Vernier LabPro or CBL2 interface. If you are using the Vernier direct-connect temperature probe, you will need a DIN-BTA (formerly CBL-DIN) adapter to convert from the 5-pin Din connector to the BTA connector.

SAFETY CAUTIONS

1. Because several of these liquids are highly volatile, keep the room well ventilated. Cap the test tubes or bottles at times when the experiment is not being performed. The experiment should not be performed near any open flames.

2. When using chemicals, students should wear aprons, gloves, and goggles.

Graphing Calculator and Sensors

TIPS AND TRICKS

- Students should have the DataMate program loaded on their graphing calculators. Refer to Appendix B of Vernier's *Chemistry with Calculators* for instructions.

- Not all models of TI graphing calculators have the same amount of memory. If possible, instruct students to clear all calculator memory before loading the DataMate program.

TECHNIQUES TO DEMONSTRATE

When viewing graphs on the calculator, students should use the arrow keys to trace the data points on the graph.

If students are using only a single temperature probe and would like to see two or three substances on the same graph, instruct them to store the first and/or second data set before beginning the next substance. From the Main Screen of DataMate, the Store Latest Run feature can be found under the Tools menu. The program will permit storing only up to two runs. If more than one sensor is used at a time, the Store Latest Run feature will not work.

Experimental Setup

TIPS AND TRICKS

- Perform this experiment in a fume hood or in a well-ventilated classroom.

- Other liquids can be substituted. Although it has a somewhat larger ΔT, 2-propanol can be substituted for 1-propanol. Some petroleum ethers have a high percentage of hexane and can be used in place of propanols. Other alkanes of relatively high purity, such as *n*-heptane or *n*-octane can also be used. Water, with a ΔT value of about 5°C, emphasizes the effect of hydrogen bonding on a low–molecular weight liquid. However, students might have difficulty comparing its hydrogen bonding capability with that of the alcohols used.

- Other properties, besides ΔT values, vary with molecular size and consequent size of intermolecular forces of attraction. Viscosity increases noticeably from methanol through 1-butanol. The boiling temperatures of methanol, ethanol, 1-propanol, and 1-butanol are 65°C, 78°C, 97°C, and 117°C, respectively.

Evaporation and Intermolecular Attractions *continued*

Answers
ANALYSIS

1. Of the alcohols, methanol evaporated the fastest and had the largest ΔT value. The molecular weight of methanol is 32.

2. Of the alcohols, 1-butanol evaporated the slowest and had the smallest ΔT value. The molecular weight of 1-butanol is 74.

3. The substances n-pentane and 1-butanol have similar molecular weights of 72 and 74. In addition to dispersion forces, 1-butanol possesses hydrogen bonding between its molecules. This results in a stronger attraction and a slower rate of evaporation, which results in a smaller ΔT value.

4. Dispersion forces and hydrogen bonding.

5. Graph of ΔT *vs.* molecular weight.

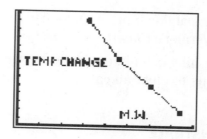

CONCLUSIONS

1. Ethylene glycol has two –OH groups, whereas 1-propanol has only one. This additional –OH group increases the amount of hydrogen bonding resulting in a significantly smaller ΔT than 1-propanol.

2. The best ink solvent, based on rate of evaporation, would be n-pentane, which had a ΔT of 16.1°C.

Data Tables with Sample Data
DATA TABLE

Substance	T_1 (°C)	T_2 (°C)	$\Delta T (T_1 - T_2)$ (°C)	Predicted ΔT(°C)	Explanation
Ethanol	23.5	15.2	8.3		
1-propanol	23.0	18.1	4.9		
1-butanol	23.2	21.5	1.7	varies (< 4.9°C)	higher molecular weight than 1-propanol (both have H-bonds)
n-pentane	23.0	6.9	16.1	varies (> 8.3°C)	higher molecular weight than either, but no H-bonding
Methanol	22.9	9.8	13.1	varies (> 8.3°C)	lower molecular weight than ethanol (both have H-bonds)
n-hexane	23.2	11.2	12.0	varies (< 16.1°C)	higher molecular weight than n-pentane; also no H-bonding

PRE-LAB RESULTS

Substance	Formula	Structural formulas	Molecular weight	Hydrogen bond (yes or no)

Ethanol	C_2H_5OH	H H
		  | |
		H—C—C—O—H
		  | |
		  H H
			46	yes

1-propanol	C_3H_7OH	H H H
		  | | |
		H—C—C—C—O—H
		  | | |
		  H H H
			60	yes

1-butanol	C_4H_9OH	H H H H
		  | | | |
		H—C—C—C—C—O—H
		  | | | |
		  H H H H
			74	yes

n-pentane	C_5H_{12}	H H H H H
		  | | | | |
		H—C—C—C—C—C—H
		  | | | | |
		  H H H H H
			72	no

Methanol	CH_3OH	H | H—C—O—H | H	32	yes

n-hexane	C_6H_{14}	H H H H H H
		  | | | | | |
		H—C—C—C—C—C—C—H
		  | | | | | |
		  H H H H H H
			86	no

Skills Practice Lab

Evaporation and Intermolecular Attractions

You are an organic chemist working for a company that manufactures various types of ink. You have been asked to create a calligrapher's ink that dries quickly at room temperature. The company feels that such a product would be a big hit since faster-drying ink would cause less distortion to the paper on which it is used. Ink consists of two components: a pigment, or coloring agent, and a solvent. The pigment is what gives the ink its color. The solvent is the chemical in which the pigment is dissolved. Your job is to select a solvent that will evaporate quickly.

To determine the best solvent to use, you will test two types of organic compounds—alkanes and alcohols. To establish how and why these substances evaporate, you will test four alcohols and two alkanes. From your results, you will be able to predict how other alcohols and alkanes will evaporate.

The two alkanes you will test are pentane, C_5H_{12}, and hexane, C_6H_{14}. Alkanes contain only carbon and hydrogen atoms, while alcohols also contain the –OH functional group. In this experiment, two of the alcohols you will test are methanol, CH_3OH, and ethanol, C_2H_5OH. To better understand why these substances evaporate, you will examine the molecular structure of each for the presence and relative strength of hydrogen bonding and London dispersion forces.

The process of evaporation requires energy to overcome the intermolecular forces of attraction. For example, when you perspire on a hot day, the water molecules in your perspiration absorb heat from your body and evaporate. The result is a lowering of your skin temperature known as evaporative cooling.

In this experiment, temperature probes will be placed into small containers of your test substances. When the probes are removed, the liquid on the temperature probes will evaporate. The temperature probes will monitor the temperature change. Using your data, you will determine the temperature change, ΔT, for each substance and relate that information to the substance's molecular structure and presence of intermolecular forces.

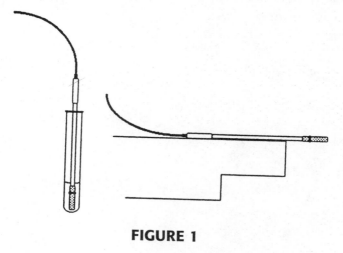

FIGURE 1

Evaporation and Intermolecular Attractions *continued*

OBJECTIVES

- **Measure** temperature changes.
- **Calculate** changes in temperature.
- **Relate** temperature changes to molecular bonding.
- **Predict** temperature changes for various liquids.

MATERIALS

- 1-butanol
- ethanol (ethyl alcohol)
- *n*-hexane
- *n*-pentane
- 1-propanol

- methanol (methyl alcohol)
- filter paper pieces, 2.5 cm × 2.5 cm (6)
- masking tape
- rubber bands, small (2)

EQUIPMENT

- LabPro or CBL2 interface
- TI graphing calculator
- Vernier temperature probes (2)

SAFETY

- Wear safety goggles when working around chemicals, acids, bases, flames, or heating devices. Contents under pressure may become projectiles and cause serious injury.
- If any substance gets in your eyes, notify your instructor immediately and flush your eyes with running water for at least 15 minutes.
- Use flammable liquids only in small amounts.
- When working with flammable liquids, be sure that no one else in the lab is using a lit Bunsen burner or plans to use one. Make sure there are no other heat sources present.
- Secure loose clothing, and remove dangling jewelry. Do not wear open-toed shoes or sandals in the lab.
- Wear an apron or lab coat to protect your clothing when working with chemicals.
- Never return unused chemicals to the original container; follow instructions for proper disposal.
- Always use caution when working with chemicals.
- Never mix chemicals unless specifically directed to do so.
- Never taste, touch, or smell chemicals unless specifically directed to do so.

Name _____ Class _____ Date _____

Evaporation and Intermolecular Attractions *continued*

Pre-lab Procedure

Before doing the experiment, complete the pre-lab table. The name and formula are given for each compound. Draw a structural formula for a molecule of each compound. Then determine the molecular weight of each of the molecules. Dispersion forces exist between any two molecules and generally increase as the molecular weight of the molecule increases. Next, examine each molecule for the presence of hydrogen bonding. Before hydrogen bonding can occur, a hydrogen atom must be bonded directly to an N, O, or F atom within the molecule. Tell whether or not each molecule has hydrogen-bonding capability.

Substance	Formula	Structural formulas	Molecular weight	Hydrogen bond (yes or no)
Ethanol	C_2H_5OH			
1-propanol	C_3H_7OH			
1-butanol	C_4H_9OH			
n-pentane	C_5H_{12}			
Methanol	CH_3OH			
n-hexane	C_6H_{14}			

Procedure

EQUIPMENT PREPARATION

1. Obtain and wear goggles! **CAUTION:** *The compounds used in this experiment are flammable and poisonous. Avoid inhaling their vapors. Avoid their contact with your skin or clothing. Be sure there are no open flames in the lab during this experiment. Notify your teacher immediately if an accident occurs.*

2. Plug temperature probe 1 into Channel 1 and temperature probe 2 into Channel 2 of the LabPro or CBL 2 interface. Use the link cable to connect the TI graphing calculator to the interface. Firmly press in the cable ends.

3. Turn on the calculator, and start the DATAMATE program. Press [CLEAR] to reset the program.

4. Set up the calculator and interface for two temperature probes.

   a. Select SETUP from the main screen.

   b. If the calculator displays two temperature probes, one in CH 1 and another in CH 2, proceed directly to Step 5. If it does not, continue with this step to set up your sensor manually.

   c. Press [ENTER] to select CH 1.

   d. Select TEMPERATURE from the SELECT SENSOR menu.

   e. Select the temperature probe you are using (in °C) from the TEMPERATURE menu.

Evaporation and Intermolecular Attractions *continued*

f. Press ▼ once, then press ENTER to select CH2.

g. Select TEMPERATURE from the SELECT SENSOR menu.

h. Select the temperature probe you are using (in °C) from the TEMPERATURE menu.

5. Set up the data-collection mode.

a. To select MODE, use ▲ to move the cursor to MODE and press ENTER.

b. Select TIME GRAPH from the SELECT MODE menu.

c. Select CHANGE TIME SETTINGS from the TIME GRAPH SETTINGS menu.

d. Enter "3" as the time between samples in seconds.

e. Enter "80" as the number of samples. (The length of the data collection will be four minutes.)

f. Select OK to return to the setup screen.

g. Select OK again to return to the main screen.

6. Wrap probe 1 and probe 2 with square pieces of filter paper secured by small rubber bands as shown in **Figure 1.** Roll the filter paper around the probe tip in the shape of a cylinder. **Hint:** First slip the rubber band up on the probe, wrap the paper around the probe, and then finally slip the rubber band over the wrapped paper. The paper should be even with the probe end.

7. Stand probe 1 in the ethanol container and probe 2 in the 1-propanol container. Make sure the containers do not tip over.

8. Prepare two pieces of masking tape, each about 10 cm long, to be used to tape the probes in position during Step 9.

DATA COLLECTION

9. After the probes have been in the liquids for at least 30 seconds, select START to begin collecting temperature data. A live graph of temperature versus time for both Probe 1 and probe 2 is being plotted on the calculator screen. The live readings are displayed in the upper-right corner of the graph, the reading for probe 1 first, the reading for probe 2 below it. Monitor the temperature for 15 seconds to establish the initial temperature of each liquid. Then simultaneously remove the probes from the liquids, and tape them so the probe tips extend 5 cm over the edge of the table top as shown in **Figure 1.**

10. Data collection will stop after four minutes (or press the STO▶ key to stop *before* four minutes have elapsed). On the displayed graph of temperature versus time, each point for probe 1 is plotted with a dot, and each point for probe 2 with a box. As you move the cursor right or left, the time (X) and temperature (Y) values of each probe 1 data point are displayed below the graph. Based on your data, determine the maximum temperature, T_1, and minimum temperature, T_2. Record T_1 and T_2 for probe 1.

Press ▼ to switch the cursor to the curve of temperature versus time for probe 2. Examine the data points along the curve. Record T_1 and T_2 for probe 2.

11. For each liquid, subtract the minimum temperature from the maximum temperature to determine ΔT, the temperature change during evaporation.

12. Roll the rubber band up the probe shaft, and dispose of the filter paper as directed by your instructor.

13. Based on the ΔT values you obtained for these two substances, plus information in the pre-lab exercise, *predict* the ΔT value for 1-butanol. Compare its hydrogen-bonding capability and molecular weight with those of ethanol and 1-propanol. Record your predicted ΔT, and then explain how you arrived at this answer in the space provided. Do the same for *n*-pentane. It is not important that you predict the exact ΔT value; simply estimate a logical value that is higher, lower, or between the previous ΔT values.

14. Press ENTER to return to the main screen. Test your prediction in Step 13 by repeating Steps 6–12 using 1-butanol with probe 1 and *n*-pentane with probe 2.

15. Based on the ΔT values you have obtained for all four substances, plus information in the pre-lab exercise, predict the ΔT values for methanol and *n*-hexane. Compare the hydrogen-bonding capability and molecular weight of methanol and *n*-hexane with those of the previous four liquids. Record your predicted ΔT, and then explain how you arrived at this answer in the space provided.

16. Press ENTER to return to the main screen. Test your prediction in Step 15 by repeating Steps 6–12, using methanol with probe 1 and *n*-hexane with probe 2.

DATA TABLE

Substance	T_1 (°C)	T_2 (°C)	$\Delta T \ (T_1 - T_2)$ (°C)		
Ethanol					
1-propanol				**Predicted** ΔT(°C)	**Explanation**
1-butanol					
n-pentane					
Methanol					
n-hexane					

Name _____ Class _____ Date _____

Evaporation and Intermolecular Attractions *continued*

Analysis

1. Analyzing data Which of the tested alcohols evaporated the fastest? Which alcohol had the largest ΔT value? What was the alcohol's molecular weight?

_____

_____

_____

2. Analyzing data Which of the alcohols tested evaporated the slowest? Which alcohol had the smallest ΔT value and molecular weight? _____

_____

_____

_____

3. Analyzing results The alcohol 1-butanol and alkane *n*-pentane have similar molecular weights, but their tests resulted in very different ΔT values. Based on the information in your pre-lab data table, explain the difference in the ΔT values of these substances. _____

_____

_____

_____

_____

4. Analyzing information What types of intermolecular forces are evident in this experiment? _____

_____

_____

5. Analyzing data Make a graph of your data with the molecular weight of each substance on the *x*-axis and ΔT on the *y*-axis.

Evaporation and Intermolecular Attractions *continued*

Conclusions

1. Evaluating results Alcohols with more than one –OH group are known as glycols. The substance ethylene glycol, CH_2OHCH_2OH, has a molecular weight of 62. Based on your data, would you expect it to have a larger or smaller ΔT than 1-propanol? Explain your answer using the results of this experiment.

_____

_____

_____

_____

_____

2. Inferring conclusions Of the substances tested in this experiment, which would work best as the solvent for the ink your company is developing?

_____

_____

Extensions

1. Applying results Using methanol and ethanol as solvent choices, test different types of mediums. Use the different medium choices in place of the filter paper to determine if the medium has any effect on the rate of evaporation.

Answer Key

Concept Review: States and State Changes

1. solid, melting, melting point
2. liquid, viscous
3. cohesion, adhesion, surface tension, evaporation
4. gas, boiling point, condensation
5. freezing, freezing point
6. sublimation, deposition

Concept Review: Intermolecular Forces

1. Ionic substances consist of separate ions, each of which is attracted not just to one ion of opposite charge, but to all ions of opposite charge in its vicinity. This attraction tends to hold the substance together until high temperatures are reached.
2. Substances with weak intermolecular forces must be cooled to low temperatures before there is enough attraction between molecules to hold the molecules together in the solid state.
3. The molecules of substances with strong intermolecular forces are attracted to each other and form intermolecular bonds that hold the molecules together in a solid state.
4. Dipole forces tend to hold neighboring molecules together, requiring higher temperatures to reach the melting and boiling points.
5. In a hydrogen bond, an exposed hydrogen proton is attracted to an adjacent atom or group of atoms with a high electronegativity.
6. Hydrogen bonding is responsible for the formation of relatively strong bonds between molecules, resulting in a relatively high boiling point, greater surface tension, and other properties characteristic of water.
7. A molecule can become an momentary dipole when its electrons become unequally distributed around its nuclei.

8. London forces
9. London forces and dipole forces are usually much weaker than forces between ions.
10. The larger particles are, the farther apart they are and the smaller the effects of the attraction are. Molecular shape may also affect attractive forces. For example, if molecules are large but have a flat shape, they can come close together and attractive forces have a greater effect.

Concept Review: Energy of State Changes

1. molar enthalpy of fusion
2. molar enthalpy of vaporization
3. molar heat capacities
4. hydrogen bonding
5. 359 K
6. more
7. absorbed
8. greater
9. higher, decrease
10. oppose
11. well below
12. endothermic
13. spontaneity
14. negative
15. decrease
16. equilibrium
17. divided
18. disorder
19. condensation
20. a large

Concept Review: Phase Equilibrium

1. phase, equilibrium, vapor pressure, normal boiling point
2. phase diagram, triple point, critical point, supercritical fluid
3. temperature
4. The fraction of very energetic particles that can escape approximately doubles or triples for a 10°C increase in temperature.

Holt Chemistry

States of Matter and Intermolecular Forces

5. 0.01°C, 0.6 kPa, triple point
   100°C, 101 kPa, normal boiling point
   0°C, 101 kPa, normal freezing/melting point
   374°C, 22 MPa, critical point
6. solid, liquid, vapor
7. solid, vapor; sublimation
8. Line segment CA, negative slope, melting point decreases

Answer Key

Quiz—Section: States and State Changes

1. a
2. d
3. c
4. b
5. d
6. c
7. a
8. c
9. a
10. b

Quiz—Section: Intermolecular Forces

1. a
2. b
3. c
4. c
5. a
6. b
7. b
8. a
9. d
10. d

Quiz—Section: Energy of State Changes

1. b
2. a
3. a
4. b
5. c
6. d
7. b
8. c
9. a
10. b

Quiz—Section: Phase Equilibrium

1. c
2. b
3. d
4. b
5. b
6. a
7. b
8. a
9. d
10. c

Chapter Test

1. b
2. a
3. b
4. c
5. c
6. c
7. d
8. b
9. a
10. d
11. b
12. a
13. a
14. c
15. c
16. a
17. c
18. a
19. a
20. d

21. Cohesion refers to the attraction of particles in a liquid for each other. The particles in a liquid attract each other, producing the smallest volume possible, resulting in surface tension. In capillary action, adhesion attracts molecules in a liquid to the surface of a solid, but cohesion pulls other liquid molecules along.

22. Point B is the melting point of water, and point C is its boiling point.

23. At point D, water is a solid. At point G, it is a gas.

24. Point E is on the solid-liquid equilibrium line. Point F is on the liquid-gas equilibrium line. If the temperature is lowered, point E will become solid, and point F will become liquid.

25. $T_{mp} = \Delta H_{fus} / \Delta S_{fus} = \dfrac{12\ 400\ \text{J/mol}}{9.27\ \text{J/mol·K}} =$ 1340 K

States of Matter and Intermolecular Forces

MULTIPLE CHOICE

1. If the particles in a sample of matter have an orderly arrangement and move only in place, the sample is a
   a. gas.
   b. liquid.
   c. solid.
   d. substance.

   Answer: C   Difficulty: I   Section: 1   Objective: 1

2. The process by which a solid becomes a liquid is
   a. freezing.
   b. melting.
   c. condensation.
   d. sublimation.

   Answer: B   Difficulty: I   Section: 1   Objective: 2

3. Which of the following has a melting point greater than room temperature?
   a. oxygen
   b. water
   c. iron
   d. bromine

   Answer: C   Difficulty: II   Section: 1   Objective: 2

4. A solid with a low melting point is most likely
   a. ionic.
   b. covalent.
   c. metallic.
   d. None of the above

   Answer: B   Difficulty: II   Section: 2   Objective: 1

5. The strength of the forces between ions depends on
   a. the size of the ion only.
   b. the charge on the ion only.
   c. both the size of the ion and its charge.
   d. neither the size of the ion nor its charge.

   Answer: C   Difficulty: I   Section: 2   Objective: 1

6. The attraction of the positive end of one molecule for the negative end of a neighboring molecule is a(n)
   a. covalent bond.
   b. dipole-dipole force.
   c. ionic bond.
   d. London dispersion force.

   Answer: B   Difficulty: I   Section: 2   Objective: 2

7. Because of dipole-dipole forces, the melting points of highly polar compounds are _____ those of less polar compounds.
   a. higher than
   b. lower than
   c. equal to
   d. not comparable to
   Answer: A   Difficulty: I   Section: 2   Objective: 2

8. A hydrogen bond is a special form of a(n)
   a. covalent bond.
   b. dipole-dipole force.
   c. ionic bond.
   d. London dispersion force.
   Answer: B   Difficulty: I   Section: 2   Objective: 3

9. A hydrogen bond forms between molecules that contain hydrogen bonded to
   a. a highly electronegative atom.
   b. an atom with a low electronegativity.
   c. another hydrogen atom.
   d. a metal.
   Answer: A   Difficulty: I   Section: 2   Objective: 3

10. London dispersion forces occur between molecules that are
    a. highly polar.
    b. slightly polar
    c. nonpolar.
    d. None of the above
    Answer: C   Difficulty: I   Section: 2   Objective: 4

11. A London dispersion force is considered a dipole-dipole force because it
    a. affects all types of compounds.
    b. affects nonpolar molecules.
    c. affects polar molecules.
    d. results from a temporary dipole.
    Answer: D   Difficulty: II   Section: 2   Objective: 4

12. The amount of energy needed to melt KCl is that compound's
    a. enthalpy of fusion.
    b. enthalpy of vaporization.
    c. entropy.
    d. free energy.
    Answer: A   Difficulty: I   Section: 3   Objective 1

13. For one mole of any substance at its boiling point, the difference in the enthalpy of its vapor and the enthalpy of its liquid is its molar
    a. entropy of vaporization.
    b. enthalpy of vaporization.
    c. entropy of fusion.
    d. enthalpy of fusion.
    Answer: B   Difficulty: II   Section: 3   Objective: 1

14. Natural events are often accompanied by a(n)
    a. decrease in energy and a decrease in disorder.
    b. decrease in energy and an increase in disorder.
    c. increase in energy and a decrease in disorder.
    d. increase in energy and an increase in disorder.

    Answer: B   Difficulty: I   Section: 3   Objective: 2

15. Enthalpy changes accompanying a change of state are _____ than those accompanying the heating of a substance at each state.
    a. much smaller   c. much larger
    b. slightly smaller   d. slightly larger

    Answer: C   Difficulty: II   Section 3   Objective: 2

16. In ΔG calculations, temperature is expressed in
    a. degrees Celsius.
    b. kelvins.
    c. degrees Fahrenheit.
    d. kilojoules.

    Answer: B   Difficulty: I   Section: 3   Objective: 3

17. Spontaneity is favored by large positive values of
    a. ΔG.
    b. ΔH.
    c. ΔS.
    d. Kelvin temperature.

    Answer: C   Difficulty: I   Section: 3   Objective: 3

18. For a system in which $\Delta H_{vap} > T\Delta S_{vap}$, what state is favored?
    a. solid
    b. liquid
    c. gas
    d. No state is favored.

    Answer: B   Difficulty: III   Section: 3   Objective: 4

19. What equation is used to find the boiling point of a substance?
    a. $\Delta H_{vap} > T\Delta S_{vap}$
    b. $\Delta H_{vap} < T\Delta S_{vap}$
    c. $T = \Delta S_{vap}/\Delta H_{vap}$
    d. $T = \Delta H_{vap}/\Delta S_{vap}$

    Answer: D   Difficulty: II   Section: 3   Objective: 4

20. Pressure has an effect on boiling point because pressure affects
    a. the entropy of a gas.
    b. the entropy of a liquid.
    c. the entropy of both a gas and a liquid.
    d. neither the entropy of a gas nor the entropy of a liquid.

    Answer: A   Difficulty: I   Section: 3   Objective: 5

21. At which of the following elevations relative to sea level will the boiling point of water be the greatest?
    a. 3.1 km
    b. sea level
    c. −123 m
    d. 127 m

    Answer: C   Difficulty: I   Section: 3   Objective: 5

22. When particles are moving between two phases and no net change in the amount of either phase occurs, _____ exist(s).
    a. equilibrium
    b. sublimation
    c. two systems
    d. vapor pressure
    Answer: A   Difficulty: I   Section: 4   Objective: 1

23. A carbonated beverage over ice contains _____ phases.
    a. one
    b. two
    c. three
    d. four
    Answer: C   Difficulty: II   Section: 4   Objective: 1

24. Vapor pressure is measured when a liquid and its vapor are
    a. in equilibrium.
    b. at standard temperature.
    c. in one phase.
    d. Both (a) and (b)
    Answer: A   Difficulty: I   Section: 4   Objective: 2

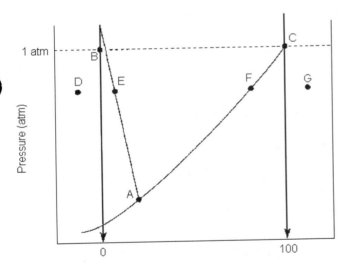

25. In the above figure, point B represents the _____ of water, and point C represents the _____ of water.
    a. normal freezing point, normal boiling point
    b. normal boiling point, normal freezing point
    c. vapor pressure, atmospheric pressure
    d. critical point of the liquid, critical point of the vapor
    Answer: A   Difficulty: III   Section: 4   Objective: 3

26. Point A in the figure represents the _____ point of water.
    a. boiling
    b. critical
    c. freezing
    d. triple

    Answer: D   Difficulty: III   Section: 4   Objective: 3

27. In the figure, water is a _____ at point D and a _____ at point G.
    a. solid, liquid
    b. liquid, gas
    c. solid, gas
    d. gas, solid

    Answer: C   Difficulty: III   Section: 4   Objective: 3

COMPLETION

28. The particles in a(n) _____ have an orderly, fixed arrangement.

    Answer: solid   Difficulty: I   Section: 1   Objective: 1

29. The tendency of liquids to decrease their surface area to the smallest size possible is called _____.

    Answer: surface tension

    Difficulty: I   Section: 1   Objective: 1

30. When a liquid is heated, the temperature stops rising at the liquid's _____.

    Answer: boiling point

    Difficulty: II   Section: 1   Objective: 2

31. When a solid sublimes, it goes from a(n) _____ to a(n) _____.

    Answer: solid, gas   Difficulty: I   Section: 1   Objective: 2

32. The type of bonding in a substance that is a liquid at room temperature will most likely be _____.

    Answer: covalent   Difficulty: II   Section: 2   Objective: 1

33. The melting point of $BaCl_2$ is _____ than the melting point of $CaCl_2$.

    Answer: lower   Difficulty: II   Section: 2   Objective: 1

34. The greater the difference in electronegativities in a diatomic molecule, the _____ the polarity of the molecule.

    Answer: greater   Difficulty: I   Section 2   Objective: 2

35. Weak dipole-dipole forces result in a(n) _____ melting point.

    Answer: low   Difficulty: II   Section: 2   Objective: 2

36. A hydrogen bond is _____ than a London dispersion force and _____ than an ionic bond.

    Answer: stronger, weaker

    Difficulty: I   Section: 2   Objective: 3

37. The base pairs on complementary strands of DNA are held together by _____.

    Answer: hydrogen bonds

    Difficulty: I   Section: 2   Objective: 3

38. If it were not for London dispersion forces, most nonpolar compounds would be _____ at room temperature.

    Answer: gases   Difficulty: I   Section: 2   Objective: 4

39. London dispersion forces are considered to be dipole-dipole forces because they cause _____ in nonpolar molecules.
    Answer: temporary dipoles
    Difficulty: II   Section: 2   Objective: 4

40. Water boils at 373 K. The molar enthalpy of vaporization for water is 40.7 kJ/mol. If one mole of water is at 373 K when it starts to boil, its temperature when it all has boiled will be _____.
    Answer: 373 K   Difficulty: II   Section: 3   Objective: 1

41. The total energy content of a system is its _____, and the measure of the system's disorder is its _____.
    Answer: enthalpy, entropy
    Difficulty: I   Section: 3   Objective: 1

42. Of steam, ice water, liquid water, and crushed ice, _____ has the highest entropy.
    Answer: steam   Difficulty: I   Section: 3   Objective: 2

43. If a process increases entropy, the process is likely to be _____.
    Answer: spontaneous   Difficulty: I   Section: 3   Objective: 2

44. If a change of state is at equilibrium, ΔH _____ $T\Delta S$.
    Answer: is equal to   Difficulty: II   Section: 3   Objective: 3

45. If Gibbs energy change is positive, a change will not occur unless _____ is added to the system.
    Answer: energy   Difficulty: I   Section: 3   Objective: 3

46. If $\Delta H_{vap} < T\Delta S_{vap}$, the _____ state is favored.
    Answer: gaseous   Difficulty: II   Section: 3   Objective: 4

47. If $\Delta S_{fus} = 10$ J/mol•K and $\Delta H_{fus} = 2000$ J/mol for a certain substance, its melting point will be _____.
    Answer: 200 K   Difficulty: II   Section: 3   Objective: 4

48. Pressure changes have no effect on the entropy of substances in the _____ or _____ states.
    Answer: liquid, solid (solid, liquid)
    Difficulty: I   Section: 3   Objective: 5

49. The instructions on a box of cake mix might give high-altitude instructions for baking the cake. The temperature given for baking the cake would be _____ than the temperature given for baking it at sea level.
    Answer: lower   Difficulty: III   Section: 3   Objective: 5

50. A region that has the same composition and properties throughout is a(n) _____.
    Answer: phase   Difficulty: I   Section: 4   Objective: 1

51. Sweetened iced tea contains _____ phases.
    Answer: two (liquid and solid)
    Difficulty: II   Section: 4   Objective: 1

52. Vapor pressure _____ when temperature decreases.
    Answer: decreases   Difficulty: I   Section: 4   Objective: 2

53. At 630 K, the vapor pressure of mercury equals atmospheric pressure, so 630 K is the _____ of mercury.
    Answer: boiling point
    Difficulty: II   Section: 4   Objective: 2

54. The equilibrium lines on a phase diagram meet at the _____.

    Answer: triple point   Difficulty: I   Section: 4   Objective: 3

55. On a phase diagram, the liquid and vapor phases of a substance are impossible to tell apart above the _____ point.

    Answer: critical   Difficulty: I   Section: 4   Objective: 3

SHORT ANSWER

56. Explain the difference between the arrangement of particles in a liquid and in a gas.

    Answer:

    Particles in a liquid are attracted to each other but are able to move past each other. Particles in a gas have little attraction for each other and are free to move anywhere within the container.

    Difficulty: II   Section: 1   Objective 1

57. Explain the difference in adhesion and cohesion in terms of capillary action.

    Answer:

    Cohesion is the attraction of water molecules for each other, and adhesion is attraction of water molecules for solid surfaces. Capillary action occurs because adhesion pulls water molecules along a solid and cohesion causes those water molecules to pull other molecules along with them.

    Difficulty: III   Section: 1   Objective: 1

58. Compare evaporation and boiling.

    Answer:

    Both processes involve changing from a liquid to a gas. Boiling occurs at the liquid's boiling point. Evaporation occurs at temperatures below the boiling point.

    Difficulty: II   Section: 1   Objective: 2

59. Compare freezing and deposition.

    Answer:

    Both processes involve formation of a solid. In freezing, the solid forms from a liquid. During deposition, the solid forms from a gas.

    Difficulty: II   Section: 1   Objective: 2

60. How do intermolecular forces differ from attraction between ions?

    Answer:

    Intermolecular forces are weaker.

    Difficulty: I   Section: 2   Objective: 1

61. Why don't dipole-dipole forces have much affect on a gas?

    Answer:

    The gas particles are too far apart.

    Difficulty: I   Section: 2   Objective: 2

62. Explain why water contains strong hydrogen bonds and H_2S does not.

    Answer:

    Oxygen is highly electronegative, and sulfur is not.

    Difficulty: I   Section: 2   Objective: 3

63. Explain how the size and shape of particles affect the attraction between them.

    Answer:

    In general, the larger the particle, the smaller the attraction because the particles are farther apart. The shape of the particle also determines how close the particles can come to each other. The closer the particles, the stronger the attraction.

    Difficulty: II   Section: 2   Objective: 4

64. Ice at −20°C is heated. The temperature rises until the ice starts to melt. Explain why the temperature doesn't rise further until all the ice has melted.
    Answer:
      The energy absorbed is used to break the attraction between the molecules in the ice crystal instead of increasing the kinetic energy of the particles.
    Difficulty: II   Section: 1   Objective: 2

65. Explain why the effects of entropy increase as temperature increases.
    Answer:
      An increase in temperature causes particles to have more kinetic energy. This effect increases the movement of the particles, contributing to less order in the substance.
    Difficulty: II   Section: 3   Objective: 2

66. How is a change in Gibbs energy related to changes in enthalpy and entropy?
    Answer:
      The change in free energy is the difference between the change in enthalpy and the product of the Kelvin temperature and the entropy change. This relationship can be stated mathematically as $\Delta G = \Delta H - T\Delta S$.
    Difficulty: II   Section: 3   Objective: 3

67. In terms of Gibbs energy, why does $\Delta H_{vap} = T\Delta S_{vap}$ at the boiling point of a substance?
    Answer:
      For these two terms to be equal, ΔG must be zero. Gibbs energy does equal zero at equilibrium. At the boiling point, vapor and liquid are at equilibrium.
    Difficulty: III   Section: 3   Objective: 4

68. Explain what happens in an equilibrium that involves a liquid and a solid.
    Answer:
      The particles in the solid go into the liquid at the same rate particles go from the liquid to the solid.
    Difficulty: I   Section: 4   Objective: 1

69. Explain why vapor pressure increases as temperature increases.
    Answer:
      As temperature increases, the kinetic energy of the particles in the liquid increases also. Because they have more kinetic energy, more particles can escape the liquid and form vapor, increasing the pressure.
    Difficulty: II   Section: 4   Objective: 2

70. What does a phase diagram show for a substance?
    Answer:
      It shows what state a substance is in under specific conditions of temperature and pressure.
    Difficulty: I   Section: 4   Objective: 3

71. What is a supercritical fluid?
    Answer:
      A supercritical fluid is the state that a substance is in when the liquid and vapor phases are indistinguishable.
    Difficulty: I   Section: 4   Objective: 3

PROBLEMS

72. For hydrogen sulfide at 188 K, ΔH = 2380 J/mol, and ΔS = 12.6 J/mol•K. Calculate the change in Gibbs energy. Will the change be spontaneous?

    Answer:

    $\Delta G = \Delta H - T\Delta S = 2380$ J/mol $- 188$ K $\times 12.6$ J/mol·K
    $= 2380$ J/mol $- 2369$ J/mol $= 11$ J/mol

    Because this value is positive, the change will not be spontaneous.

    Difficulty: II   Section: 3   Objective: 3

73. Calculate the melting point, T_{mp}, for bromine, Br_2.
    (ΔH_{fus} = 10.57 kJ/mol, ΔS_{fus} = 39.8 J/mol·K)

    Answer:

    $\Delta H_{fus} = T_{mp}\Delta S_{fus}$
    10570 J/mol $= T_{mp} \times 39.8$ J/mol·K
    266 K $= T_{mp}$

    Difficulty: II   Section: 3   Objective: 4

74. Calculate the melting point, T_{mp}, for lead, Pb. (ΔH_{fus} = 4.77 kJ/mol, ΔS_{fus} = 7.9 J/mol·K)

    Answer:

    $\Delta H_{fus} = T_{mp}\Delta S_{fus}$
    4770 J/mol $= T_{mp} \times 7.9$ J/mol·K
    604 K $= T_{mp}$

    Difficulty: II   Section: 3   Objective: 4

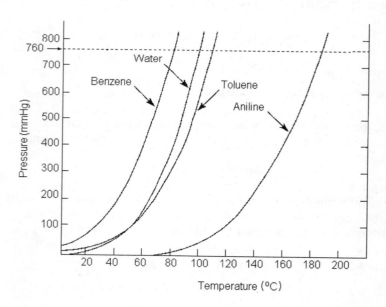

75. Which of the substances shown in the above figure has the highest boiling point? Which evaporates most easily? Explain your answers.

    Answer:

    At any particular temperature, the vapor pressure of aniline is lowest. Therefore, it must reach a higher temperature for its vapor pressure to equal atmospheric pressure. For any temperature, the vapor pressure of benzene is the highest. Therefore, it evaporates most easily.

    Difficulty: III   Section: 4   Objective: 2

Solutions Manual

Solutions for problems can also be found at go.hrw.com. Enter the keyword HW4INTTNS to obtain solutions.

Practice Problems A

1. Given: $\Delta H_{fus} = 108.9$ J/g
$\Delta S_{fus} = 31.6$ J/mol·K
$\Delta H_{vap} = 837$ J/g
$\Delta S_{vap} = 109.9$ J/mol·K

Unknown: T_{mp}, T_{bp}

$T_{mp} = \dfrac{\Delta H_{fus}}{\Delta S_{fus}} = \dfrac{5018 \text{ J/mol}}{31.6 \text{ J/mol·K}} = 159$ K

$T_{bp} = \dfrac{\Delta H_{vap}}{\Delta S_{vap}} = \dfrac{38600 \text{ J/mol}}{109.9 \text{ J/mol·K}} = 351$ K

2. Given: $\Delta H_{fus} = 8.62$ kJ/mol
$\Delta S_{fus} = 43.1$ J/mol·K
$\Delta H_{vap} = 24.9$ kJ/mol
$\Delta S_{vap} = 94.5$ J/mol·K

Unknown: T_{mp}, T_{bp}

$T_{mp} = \dfrac{\Delta H_{fus}}{\Delta S_{fus}} = \dfrac{8620 \text{ J/mol}}{43.1 \text{ J/mol·K}} = 200$ K

$T_{bp} = \dfrac{\Delta H_{vap}}{\Delta S_{vap}} = \dfrac{24900 \text{ J/mol}}{94.5 \text{ J/mol·K}} = 263$ K

3. Given: $\Delta H_{fus} = 5.66$ kJ/mol
$\Delta S_{fus} = 29.0$ J/mol·K
$\Delta H_{vap} = 23.3$ kJ/mol
$\Delta S_{vap} = 97.2$ J/mol

Unknown: T_{mp}, T_{bp}

$T_{mp} = \dfrac{\Delta H_{fus}}{\Delta S_{fus}} = \dfrac{5660 \text{ J/mol}}{29.0 \text{ J/mol·K}} = 195$ K

$T_{bp} = \dfrac{\Delta H_{vap}}{\Delta S_{vap}} = \dfrac{23\,300 \text{ J/mol}}{97.2 \text{ J/mol·K}} = 240$ K

Section 3 Review

7. Given: $\Delta H_{fus} = 10.57$ kJ/mol
$\Delta S_{fus} = 39.8$ J/mol·K

Unknown: T_{mp}

$T_{mp} = \dfrac{\Delta H_{fus}}{\Delta S_{fus}} = \dfrac{10\,570 \text{ J/mol}}{39.8 \text{ J/mol·K}} = 266$ K

8. Given: $\Delta H_{vap} = 30.0$ kJ/mol
$\Delta S_{vap} = 90.4$ J/mol·K

Unknown: T_{bp}

$T_{bp} = \dfrac{30\,000 \text{ J/mol}}{90.4 \text{ J/mol·K}} = 332$ K

9. Given: $\Delta H_{fus} = 167$ J/g
$\Delta S_{fus} = 45.3$ J/mol·K

Unknown: T_{mp}

$T_{mp} = \dfrac{\Delta H_{fus}}{\Delta S_{fus}} = \dfrac{10524 \text{ J/mol}}{45.3 \text{ J/mol·K}} = 233$ K

● ESSAY QUESTIONS

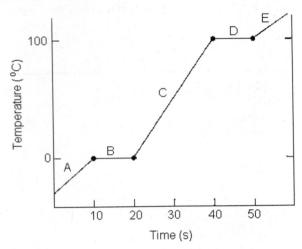

76. Explain what is happening to water in each of the sections of the temperature graph.

    Answer:
    Segment **A** shows that the energy added to ice/solid causes the temperature to rise. Segment **B** shows that the energy added causes the solid to melt, while the temperature remains constant. Segment **C** shows that the energy added increases the temperature of the liquid. Segment **D** shows that while the water boils, the temperature remains constant. Segment **E** shows that as more energy is added to the water vapor, the temperature rises.

    Difficulty: III   Section: 3   Objective: 1

77. Explain the combined effects of entropy and enthalpy during the melting of ice.

    Answer:
    The change from a solid to a liquid involves an increase in entropy because liquids are less ordered than solids. The reaction is endothermic, requiring energy from the surroundings. Therefore, the entropy change dominates over the enthalpy change.

    Difficulty: II   Section: 3   Objective: 2

78. Describe how the forces of enthalpy or entropy determine the spontaneity of a change.

    Answer:
    A spontaneous change generally involves a decrease in enthalpy. An increase in entropy often accompanies a spontaneous change. When only one of these conditions is true, spontaneity is determined by the equation $\Delta G = \Delta H - T\Delta S$.

    Difficulty: II   Section: 3   Objective: 3

79. One reason astronauts wear space suits in space is because of the decreased pressure in space. Explain what might happen to the blood of an astronaut subjected to extremely low pressures.

    Answer:
    Because blood is liquid, its boiling point is affected by pressure changes. Extremely low pressure would lower the boiling point enough that the blood would boil at a temperature less than body temperature.

    Difficulty: III   Section: 3   Objective: 5